I0043636

William Ogilvie

A Handbook to the Game-Birds

Vol. II: Pheasants (Continued), Megapodes, Curassows, Hoatzins, Bustard-Quails

William Ogilvie

A Handbook to the Game-Birds
Vol. II: Pheasants (Continued), Megapodes, Curassows, Hoatzins, Bustard-Quails

ISBN/EAN: 9783744734431

Printed in Europe, USA, Canada, Australia, Japan

Cover: Foto ©berggeist007 / pixelio.de

More available books at **www.hansebooks.com**

PLATE XXII.

RING-NECKED PHEASANT.

Wyman & Sons Limited

LLOYD'S NATURAL HISTORY.

EDITED BY R. BOWDLER SHARPE, LL.D., F.L.S., &c.

A HAND-BOOK

TO THE

GAME-BIRDS.

BY

W. R. OGILVIE-GRANT,

ZOOLOGICAL DEPARTMENT, BRITISH MUSEUM.

VOL. II.

PHEASANTS (*Continued*), MEGAPODES, CURASSOWS,
HOATZINS, BUSTARD-QUAILS.

LONDON:

EDWARD LLOYD, LIMITED,

12, SALISBURY SQUARE, FLEET STREET.

1897.

PRINTED BY

WYMAN AND SONS, LIMITED.

PREFACE.

I THINK that there can be no question as to the value of Mr. Ogilvie-Grant's volumes on the Game-Birds, and I can testify to the care which he has bestowed on the work. His volumes contain the names of every species of Game-Birds known up to the present date, so that they may be considered in the light of a small Monograph of the *Gallinæ*.

R. BOWDLER SHARPE.

Oct. 6th, 1896.

AUTHOR'S PREFACE.

THE second volume of this work contains an account of all the remaining species of the Order Gallinæ or True Game-Birds, as well as that curious and aberrant form the Hoatzin, and the Bustard-Quails.

The subject has been treated in exactly the same way as in the previous volume, and the short descriptions of the adult male and female have in almost every instance been carefully compared and revised with the specimens in the British Museum, where the collection of Game-Birds is unusually fine and perfect. Only nine of the species mentioned in the first volume are not represented in the above collection, six of the desiderata being Francolins, while of the 221 species treated of in the present volume, though a like number are wanting, two or three of these appear to be doubtfully distinct species.

Two birds are described for the first time in this volume: Whitehead's Bustard-Quail (*Turnix whiteheadi*), an extremely interesting and distinct new form met with in the neighbourhood of Manila, and Cholmley's See-See Partridge, from the Soudan and African shores of the Red Sea, a description of them both being given in the Appendix.

W. R. OGILVIE-GRANT.

SYSTEMATIC INDEX.

SYSTEMATIC INDEX.

PAGE

FAMILY CRACIDÆ. 200

I. CRAX, Linn. 200
   1. alector, Linn. 201
   2. fasciolata, Spix. 202
   3. pinima, Pelz. 202
   4. globicera, Linn.... 203
   5. panamensis, Ogilvie-Grant. 205
   6. hecki, Reichenow. 205
   7. grayi, Ogilvie-Grant. 206
   8. carunculata, Temm. 206
   9. globulosa, Spix. 208
   10. daubentoni, Gray. 209
   11. alberti, Fraser. 211

   NOTHOCRAX, Burmeister. 211
   1. urumutum (Spix). 211

II. MITUA, Strickl. 213
   1. mitu (Linn.). 214
   2. tomentosa (Spix). 215
   4. salvini, Reinhardt. 215

III. PAUXIS, Scl. 216
   1. pauxi (Linn.). 217

IV. OREOPHASIS, G. R. Gray. 217
   1. derbianus, Gray. 218

V. PENELOPE, Merrem. 220
   1. superciliaris, Temm. 220
   2. montagnii (Bonap.). 221
   3. sclateri, G. R. Gray. 222
   4. jacupeba, Spix. 223
   5. marail, Gmel. 223
   6. ortoni, Salvin. 224
   7. purpurascens, Wagler. 224
   8. obscura, Temm. 225
   9. cristata (Linn.). 226
   10. boliviana, Bonap. 227
   11. pileata, Wagler.... 228
   12. ochrogaster, Pelz. 229
   13. jacucaca, Spix 229
   14. argyrotis (Bonap.). 230
   15. albipennis, Taczanowski· 230

LIST OF PLATES.

7

GAME-BIRDS.

FAMILY PHASIANIDÆ (Vol. I. p. 78).

SUB-FAMILY PHASIANINÆ (Vol. I. p. 199).

THE CHEER PHEASANTS. GENUS CATREUS.

Catreus, Cabanis, in Ersch. und Grueb. Encycl. sec. 1, liii. p. 221 (1851).

Type, *C. wallichi* (Hardwicke).

Tail long, wedge-shaped, and composed of *eighteen* feathers, the middle pair being very long, and about five times as long as the outer pair.

First primary flight-feather shorter than the tenth; fifth slightly the longest.

A large naked patch of crimson red skin surrounding the eye.

A long full occipital crest in the male, less marked in the female. The former is also provided with a pair of strong spurs, which are sometimes represented in the latter by a pair of blunt knobs.

1. THE CHEER PHEASANT. CATREUS WALLICHI.

Phasianus wallichii, Hardwicke, Tr. Linn. Soc. xv. p. 166 (1827); Mitchell, P. Z. S. 1858, p. 544, pl. 147, fig. 1, and pl. 149, fig. 4; Elliot, Monogr. Phasian. ii. pl. x. (1872); Hume and Marshall, Game Birds of India, i. p. 169, pl. (1878); Oates, ed. Hume's Nests and Eggs, Ind. B. iii. p. 412 (1890).

Phasianus staceii, Vig. Phil. Mag. 1831, p. 232; id. P. Z. S. 1831, p. 35.

Catreus wallichii, Gould. Cent. B. Himal. pl. 68 (1832); id. Birds of Asia, vii. pl. 18 (1865); Ogilvie-Grant, Cat. B. Brit. Mus. xxii. p. 317 (1893).

Adult Male.—Top of the head dark brown, all the feathers, especially those of the crest, tipped with dirty white; throat, neck, and chest dirty white; upper back, breast, and sides of the belly creamy-white, the feathers of the former widely fringed with whitish-grey; wings ochraceous-buff, all these parts being barred and marked with black; lower back and rump *rust-colour*, with black bars glossed with green; *middle of belly black;* quills brownish-black, *edged and mottled* with buff; tail-feathers whitish-buff, the middle pair with wide irregular black bars changing to dark chestnut on the inner webs of the outer pairs. Total length, 34 inches; wing, 10; tail, 20·5; tarsus, 2·9.

Adult Female.—Differs from the male chiefly in having the feathers of the head and crest *edged with buff;* the upper back *pale chestnut,* widely barred with black; and the lower back and rump *dirty brown,* mixed with black and buff; the neck and chest black, edged with buff, the breast and belly *rufous-chestnut,* edged with buff and more or less mottled with black, the quills *regularly barred* with *buff* on the *outer* and *pale chestnut* on the *inner* webs, and the tail-feathers *brown* or *reddish-brown* with wide mottled bars of black and buff, except on the outermost pairs, which are mostly buff. Total length, 30 inches; wing, 8·9; tail, 15·5; tarsus, 2·6.

Range.—The Himalayas, extending eastwards as far as Katmandu, in Nepal, and westwards to Chamba, but not apparently to Kashmir.

Habits.—Mr. A. O. Hume writes:—" The Cheer is extremely locally distributed, and seems to me very capricious in its choice of habitations; on one side of a river you meet with plenty in suitable spots, on the other side you may search fifty square miles of most likely-looking country and never see one.

" From six to seven thousand feet is the elevation at which, in October, they are most common, but in winter and spring they go lower, and some even breed lower, and in summer they *may* be met with up to at least ten thousand feet (I myself killed a pair of old ones late in June at fully this elevation), and probably higher. Of course they are birds of the outer or wooded hills, and when once you cross a high snowy ridge, which effectually arrests the clouds of the monsoon, into dry, more or less treeless regions, like Lahoul, Spiti, and Ladákh, you lose the Cheer and all the Pheasants but the Snow-Cocks. The former are all more or less birds of the forest, and all belong to the zone of abundant rainfall.

" The best places in which to find Cheer are the Dangs, or precipitous places, so common in many parts of the interior ; not vast bare cliffs, but a whole congeries of little cliffs one above the other, each perhaps from fifteen to thirty feet high broken up by ledges, on which a man could barely walk, but thickly set with grass and bushes, and out of which grow up stunted trees, and from which hang down curious skeins of grey roots and mighty garlands of creepers.

" If the hill above be thinly wooded, and on some plateau below there are a good number of Millet and Prince's Feather fields, you are, in a Cheer district, next to certain in the autumn to find a covey on the upper ledges of such a spot, about ten o'clock in the morning. . . ."

In describing their flight, he says :—" The force with which Cheer descend is almost incredible. Other Pheasants, in descending, keep the wings a little open ; these birds pass one at such a fearful pace that it is impossible to be certain, but it always appeared to me that Cheer quite closed their wings, and I attribute their power to do this to their enormous tails sufficing to guide them. When within a hundred feet (I speak by guess), of the level at which they intend to alight, suddenly out go the wings, the tan is spread to its fullest expanse, the bird looks double the size it did a second before, and sweeps off in graceful curves right or left, shortly dropping suddenly, almost as if shot, into some patch of low cover. If no shots have been fired, you may walk straight down, and ten to one find him exactly where you marked him. . . ."

Mr. Wilson, in his excellent account of this species, tells us :—"They wander a good deal about the particular hill they are located on, but not beyond certain boundaries, remaining about one spot for several days or weeks, and then shifting to another, but never entirely abandoning the place, and year after year they may, to a certainty, be found in some quarter of it.

"During the day, unless dark and cloudy, they keep concealed in the grass and bushes, coming out, morning and evening, to feed.

"After concealing themselves, they lie very close, and are flushed within a few yards. There is, perhaps, no bird of its size which is so difficult to find after the flock have been disturbed and they have concealed themselves ; where the grass is very long, even if marked down, without a good dog, it is often impossible to flush them, and even with the assistance of the best dogs not one-half will be found a second time. A person may walk within a yard of one and it will not move. I have knocked them over with a stick, and even taken them with the hand. In autumn the long grass, so prevalent about many of the places they resort to, enables them to hide almost anywhere, but this is burnt by the villagers at the end of winter, and they then seek refuge in low jungle and brushwood, and with a dog are not so difficult to find.

"Both males and females often crow at daybreak and dusk, and in cloudy weather sometimes during the day. The crow is loud and singular, and, when there is nothing to interrupt, the sound may be heard for at least a mile. It is something like the words *chir-a-pir, chir-a-pir, chir, chir, chirwa, chirwa,* but a good deal varied ; it is often begun before complete day-light, and in spring, when the birds are numerous, it invariably ushers in the day ; in this respect it may rival the domestic cock. When pairing and scattered about, the crow is often kept up for nearly half-an-hour, first from one quarter, then another, and now and then all seem to join in a chorus. At other times it seldom lasts more than five or ten minutes.

"The Cheer Pheasant feeds chiefly on roots, for which it digs holes in the ground, grubs, insects, seeds, and berries, and, if near cultivated fields, several kinds of grain form a portion of its diet ; it does not eat grass or leaves like the rest of our Pheasants.

"It is easy to rear in confinement, and might, without diffi-culty, be naturalised in England, if it would stand the long frosts and snows of severe winters, which I imagine is rather doubtful.

"This bird flies rather heavily, and seldom very far. Like most others, it generally utters a few loud screeches on getting up, and spreads out the beautifully barred feathers of its long tail, both when flying and running. It does not perch much on trees, but will occasionally fly up into one close by, when put up by dogs. It roosts on the ground generally, and when congregated together, the whole flock huddle up in one spot. At times, however, they will roost in trees and bushes."

Hybrids.—The Cheer has been known to cross with the Common Pheasant (*Phasianus colchicus*) in confinement, and there is an example of a male hybrid in the National Collection.

Nest.—Placed in the grass among low bushes near or about the base of some precipitous hill-side at elevations of from 4,000 to 7,000 or 8,000 feet.

Eggs.—Nine to fourteen in number ; pale stone-colour or very pale whitish-buff, almost devoid of markings, but many specimens have small spots and specks of brownish-red at one or other end, while more rarely the markings are scattered all

over the surface of the shell.   Egg small for the size of the bird,
and shaped like an ordinary hen's egg.   Average measurements,
2·1 by 1·5 inches.

THE TRUE PHEASANTS. GENUS PHASIANUS.

Phasianus, Linn. S. N. i. p. 270 (1766).

Type, *P. colchicus*, Linn.

Tail composed of *eighteen* feathers, long and wedge-shaped,
the middle pair being very much longer than the outer pair.

First primary flight-feather about equal to the eighth, and
*considerably longer than the tenth.**

The male has the sides of the head covered with naked
scarlet skin ; there is no crest, but the ear-tufts are considerably
lengthened, and the feet are armed with a pair of spurs.

. The various species and sub-species of this lovely group are
all natives of Asia, extending to Japan and Formosa, and, to
facilitate their identification, may be conveniently divided into
the following groups :—

I. Crown of the head green or greenish-bronze.
  A. General colour of the lower back, rump, and upper
    tail-coverts maroon or red-bronze, glossed with
    purple or green.
      a. With *no* white ring round the neck, or
        with only traces of one (species 1 to 7,
        pp. 9-22).

* See vol. i. p. 78, and footnote to p. 199.   To these remarks it may
be added, and this is noteworthy, that the Partridge type of wing is most
perfectly developed in such birds as the Quails (*Coturnix*), capable of
long and protracted flights, while in the Argus Pheasants (*Argusianus*),
which hardly ever fly, we find the most perfect type of Pheasant-wing.
In other words, the Phasianidæ, with the outer primary quills longest,
are all birds in which great powers of flight are necessary, while those
with the inner primary quills longest, seldom use their wings, if they can
avoid doing so.   We may thus infer that the powers of flight, among
these game-birds at least, are almost entirely dependent on the shape
of the wing, and probably the same rule will be found to apply to other
groups of birds.

> *b.* With a broad white ring round the neck (species 8, pp. 22–24).

B. General colour of the lower back, rump, and upper tail-coverts greenish- or bluish-slate colour, with a rust-coloured patch on each side (except in *P. versicolor*).

> > *c.* With a white ring round the neck (species 9 to 11, pp. 24–28).
> > *d.* With no white ring round the neck, or with only traces of one (species 12 to 16, pp. 28–33).

II. Crown of the head reddish-brown (species 17, pp. 34-37).

III. Crown of the head white (species 18, pp. 37–41).

Both Sœmmerring's and Reeves's Pheasants represent peculiar and somewhat aberrant types, the latter being placed by some authors in the distinct genus *Syrmaticus*, on account of its immensely long tail. Exclusive of these two birds and the rather distinct green-bellied form (*P. versicolor*) from Japan, the remaining representatives of the genus *Phasianus* are all birds of the same type as the Common Pheasant (*P. colchicus*). It is interesting to note that, roughly speaking, all the species with white rings round their necks are northern forms, while those without this ornament, or with only traces of it, are only met with farther south. On the other hand, all the maroon-rumped species are found west of about 90° E. long., while those with grey rumps are only found to the east of this line. *P. colchicus* and its allies, which have no white ring, but have a maroon-coloured rump, inhabit the area south of about 41° N. lat., and west of about 80° E. long., while *P. mongolicus* and *P. semitorquatus*, both of which have well-marked white collars, as well as maroon rumps, are met with north of about 41°, and west of about 90° E. long. Again, among the grey-rumped species, *P. elegans* and the allied forms without white collars, or with only traces of this ornament, occur south of about 35° N. lat. and east of about 90° E. long., while of the ringed form of this section, *P. torquatus*, though it is found in China as far south as Canton (and a slightly different form occurs in Formosa), ranges far north to the Lower Amoor, and a paler represen- tative is met with to the north of the Nan-shan Mountains, which lie north of 35° N. lat. So, on the whole, we may regard the ringed form as the northern, and those with- out a ring, as the southern type ; and it seems reasonable to suppose that all the species have probably sprung from an ancestral ringed form of northern origin, and that the occur- rence of a partial white ring in certain individuals of the southern species, which are normally devoid of this ornament, is due to the fact that they still occasionally revert to the original stock. It is quite wrong to regard such partially ringed individuals as hybrids, for in most cases the country which each species inhabits, is effectually shut off by enormous ranges of mountains, which completely bar all intercourse between the ringed species and their southern allies. For example, it is not uncommon to find examples of Shaw's Pheasant (*P. shawi*) from the valleys of Yarkand with an imperfect white ring round the neck ; and it is practically impossible for this species to meet with any of the ring-necked forms, *P. semitorquatus*, from Dzungaria, being apparently shut off by high mountains and impassable deserts. It must, however, be added that there can be no doubt that *P. mongolicus*, which ranges along the valley of the Syr-Darya, does occasionally interbreed with *P. chrysomelas* from the valley of the Amu-Darya, for we have seen a wild hybrid shot at Nukus which is beyond doubt a cross between the two forms. In this instance, however, there is nothing to prevent the ranges of the two species from coalescing, and no doubt the ring-necked *P. mongolicus* occa- sionally finds its way south along the Eastern shore of the Aral. Any two species of this genus will interbreed freely

with one another in captivity, or when they have been arti-
ficially brought into the same neighbourhood, and in the
majority of cases at least the hybrids produced are perfectly
fertile. In this country it is now almost impossible to find
quite pure-bred examples of what is commonly called "the
old English Pheasant" (*P. colchicus*), for the Chinese ringed
species (*P. torquatus*), which was subsequently introduced, has
crossed with it everywhere, and almost all the birds now met
with are hybrids, displaying the characters of both species in a
greater or lesser degree.

1. *Crown of the head green or greenish-bronze ; general colour
   of the lower back, rump, and upper tail-coverts maroon
   or reddish-bronze, glossed with purple or green ; with no
   white ring round the neck, or with only traces of one.*

I. THE COMMON PHEASANT. PHASIANUS COLCHICUS.

Phasianus colchicus, Linn. S. N. i. p. 271 (1766); Gould, B.
   Europe, pl. 247 (1837); id. B. Asia, vii. pl. 34 (1869);
   Elliot, Monogr. Phasian. ii. pl. 2 (1872); Dresser, B.
   Europe, vii. p. 85, pl. 469 (1879); Ogilvie Grant, Cat. B.
   Brit. Mus. xxii. p. 320 (1893).*

Phasianus colchicus septentrionalis, Lorenz, J. f. O. 1888, p.
   572.

Adult Male.†—Crown of the head bronze-green; rest of the
head and neck dark green, shading into purple on the sides
and front of the neck. Feathers of the mantle, chest, breast,
and flanks fiery orange, the former narrowly margined with
purplish-green, the latter widely edged with rich purple; those
of the upper back and scapulars mottled in the middle with
black and buff, margined by consecutive bands of buff, black
and orange-red, and tipped with purplish-lake. Lower back,
rump, and upper tail-coverts *red-maroon, glossed with purplish-*

---

* Cf. also "Fur and Feather" Series, 8vo, 1895.

† It has been found necessary to give this rather full description of pure-
bred male and female birds for purposes of comparison.

lake or oily green, according to the way the skin is held.
Most of the wing-coverts sandy-brown; middle of breast and
sides of belly *dark purplish-green ;* middle of belly and rest of
under-parts dark brown mixed with rufous.   Tail-feathers olive
down the middle, with *narrow, wide-set,* black bars, and widely
edged on each side with rufous, glossed with purplish-lake.
Total length, 37·5 inches; wing, 10·1 ; tail, 21·2 ; tarsus, 2·8.

Adult Female.—General   colour   sandy-brown, barred   with
black ; back and sides of the neck tinged with pinkish and
with metallic purple or green margins ; feathers of the mantle
and sides of the breast and flanks chestnut, with black centres
and pinkish-grey margins ; an elongate patch of white black-
tipped feathers below the eyes ; quills more coarsely barred and
mottled with buff than in the male ; tail-feathers reddish-brown
down the middle, shading into sandy-olive on the sides and
with wide irregular triple bars of black, buff, and black.   Total
length, 24·5 inches ; wing, 8·6 ; tail, 11·5 ; tarsus, 2·4.

Range.—The Common Pheasant has been introduced in most
parts of Europe, with the exception of Spain and Portugal, and
the higher latitudes of Scandinavia and Russia.   For this reason
it is difficult, if not impossible, to state accurately the limits
of its true home.   It appears, however, to be found in a wild
state in Southern Turkey, Greece, and Asia Minor as far east
as Transcaucasia, and it extends northwards to the Volga.   On
the Island of Corsica it is also met with in a wild state, and
may have been imported at some remote period; but if it is
really indigenous there, its range must formerly have extended
much farther west than the countries mentioned above.

There is no record, as far as we know, of its importation to
the British Islands, but it is mentioned in the bills-of-fare of
the Saxon kings.

Habits.—The favourite home of the Pheasant is thick covert,
woods with plenty of undergrowth, in the immediate neighbourhood of cultivated land, where in the morning and evening the
birds can come out to feed.   Oak, hazel, and fir plantations

scattered over large parks are much resorted to, for the birds seldom stray far from the shelter of the trees, and retire on the slightest approach of danger, being decidedly shy and retiring in their habits.

Most of our readers are well acquainted with the Common Pheasant in a semi-domesticated state, when it is undoubtedly polygamous, one male pairing with many females, but there seems to be good reason for believing that this habit has been acquired ; for, in a really wild state, all the evidence, though it is certainly somewhat scanty, tends to show that this, as well as the other species of *Phasianus*, are monogamous, the cock bird remaining with the female during the period of incubation, and taking part in the duties of protecting and rearing the young. In this, as in other countries, where Pheasants are reared for sport, the greater number of birds killed are cocks, and hence in the following spring there is generally a preponderance of females ; which may account for the polygamous habits of introduced birds. The males are remarkably quarrel-some in the pairing-season, fighting fiercely with one another for the different females, the more powerful birds appropriating the lion's share for their harem. When the females have laid their full complement of eggs, the male troubles his head no more about them, leaving them to undertake all the cares of rearing their family unaided. They cannot be called good mothers, for, unlike the majority of game-birds, on the approach of danger, they seek safety in flight, leaving the young to escape and hide themselves as best they can. This habit is often extremely disastrous to the brood, especially when the chicks are very small, for, on her return, the mother is apparently perfectly satisfied with finding one or two of her scattered young, and the remainder are consequently left to perish. For this reason gamekeepers are naturally anxious that the coverts, where " wild birds " are breeding, should not be disturbed during the nesting-season, and it is hardly surprising that they should treat trespassers with scant courtesy. The majority of

birds shot annually in the large preserves in this country and
in Europe, are of course reared from eggs placed under domestic
hens, who make excellent mothers to their foster-children. On
leaving her nest in the morning and evening in search of food,
the hen Pheasant is always careful to cover her eggs with dead
leaves, and she generally quits and returns to her nest on the
wing, thus avoiding as far as possible the danger of being
tracked by her enemies.

The crow of the male resembles the syllables *or-ork*, which
are often repeated several times in succession, and may be
exactly imitated by opening the mouth and drawing the breath
in sharply to the back of the throat. This call is generally to
be heard in the morning and evening, especially about sunset,
when the birds are going to roost, but during the pairing-
season it may be heard at all times of the day, and is also
given vent to when they are flushed or suddenly startled by
the report of a gun, or a clap of thunder.

There can be no doubt that if the Pheasant were not arti-
ficially reared and annually turned down in this country, it
would soon cease to exist, for, in hard winters especially, the old
birds left for stock are largely dependent on artificial feeding.
The chief food consists of grain, seeds, berries, and young
shoots, varied with insects and grubs, wire-worms being a
favourite morsel.

Pure-bred examples of *P. colchicus* are now rarely to be met
with in England, the great majority of birds being hybrids with
the Chinese Ring-Necked Pheasant (*P. torquatus*), which was
subsequently introduced.

Like the rest of its kind, the Pheasant, though it roosts and
often perches on trees, is essentially a ground bird, and a tre-
mendous runner; the old cocks, having learnt wisdom from
past experience, frequently refuse to rise at the net and face
the guns so anxiously waiting to salute them, and may be seen
running back among the beaters as fast as their legs can carry
them. The whir made in rising is loud and startling, but

When once well on the wing, the Pheasant's flight is extremely rapid, being performed by rapid and incessant beats of the rounded wings, and when coming high, down wind, the pace at which a good "rocketer" can travel is almost incredible.

During the nesting-season the hen Pheasant has numerous enemies to contend with, the most formidable being the prowling Fox, who seizes her as she sits on her nest, and the Rook and Crows, both Hooded and Carrion, who steal and suck her eggs. A curious instance of the enormous amount of damage done by Crows came under my notice in May, 1893.

With a friend, I was passing through a Scotch fir plantation forming part of a large estate in the north of Scotland, where thousands of Pheasants are annually reared and turned down. The plantation ran along about a hundred feet above the rocky sea-coast, and as we advanced along the slippery path, we found several sucked Pheasants' eggs, evidently the work of Crows, nor had we gone far before we came suddenly upon a whole family of Hooded rascals, five young and two old birds. In the course of about a quarter of a mile, we counted over a hundred empty shells which had evidently been carried to the path and there devoured. How many more might have been discovered had we searched it is impossible to say, but we saw ample evidence of the wholesale destruction which a family of Crows is capable of committing among Pheasants' eggs. Within two miles of this spot, to his shame be it said, stood a keeper's house, where a thousand young birds were being reared. This worthy informed us that the great heat and drought then prevalent was decimating his broods of young Pheasants, who were dying in scores from a disease which attacks the eyes, and from which few recover. He volunteered the information that he had not been over to the belt of fir wood "for this two months," as there was nothing there to take him so far! A little more attention to the destruction of Hooded Crows in April might have saved a hundred or two ·of strong wild-bred birds for the sport in the fall of the year.

Female Pheasants that have become barren either from age or through disease of the ovary, *generally* assume the plumage of the cock to a greater or less extent, and we have known a number of instances in which the male plumage had been so perfectly donned that it was only by the smaller size, blunt spurs, and much shorter tail, that the true sex of the individual could be ascertained. Last year I examined a hen Pheasant in *perfectly normal* plumage, but with a well-developed *sharp* spur on each leg ; this bird, on dissection, was found to have been shot in the left ovary, a No. 2 or 3 shot (!) being there embedded, which had destroyed the organ, and given rise to an ugly tumourous growth. The wound was evidently an old-standing one, but in this instance the plumage had remained normal.

The Common Pheasant not only crosses with other species of its own kind, but hybrids are occasionally produced between it and the Black Game, Domestic Fowl, and Guinea Fowl, while instances are on record of hybrids between Pheasant and Capercaillie.

Albinos and piebald birds are by no means an uncommon occurrence among our semi-domesticated birds, but no doubt much rarer among really wild individuals.

Nest.—A mere hollow in the ground, roughly lined with dead leaves, and carefully hidden from view by dead fern, brambles, or coarse grass or other herbage.

Eggs.—Vary in number from eight to twelve, but a score or more are sometimes found in one nest, probably the produce of more than one female ; they are broad ovals, slightly pointed at the smaller end, generally brown or olive-brown in colour, more rarely bluish-green, uniform in tint, and with rather a smooth polished shell. Average measurements, 1·8 by 1·4 inch.

SUB-SP. *a*. THE TALISCH PHEASANT. PHASIANUS TALISCHENSIS.

Phasianus persicus talischensis, Lorenz, J. f. O. 1888, p. 571.

Phasianus talischensis, Ogilvie-Grant, Cat. B. Brit. Mus. xxii.
  p. 324 (1893).

This is an intermediate form between *P. colchicus* and the
following species, *P. persicus*, but most nearly allied to the
former.

Adult Male.—Differs from *P. persicus* and resembles *P.
colchicus* in having the wing-coverts *sandy-brown* instead of
nearly white; on the other hand, the middle of the breast and
sides of the belly are *purplish-carmine*, and the feathers of the
chest and upper breast are narrowly margined with purple, as
in *P. persicus*.

Adult Female.—Similar to the female of *P. colchicus*.

Range.—This sub-species inhabits Talisch, a district bordering the south-western shore of the Caspian Sea. It is most
likely this bird (called *P. colchicus* by Mr. W. T. Blanford)
is plentiful throughout Mazandarán to the Gurgan River, which
enters the Caspian Sea on the south coast. To the north
of this, true *P. persicus* is found.

Habits.—No doubt perfectly similar to those of its Western
ally.

II. THE PERSIAN PHEASANT. PHASIANUS PERSICUS.

Phasianus persicus, Severtz. Bull. Mosc. xlviii. pt. 3, p. 208
  (1875); Ogilvie-Grant, Cat. B. Brit. Mus. xxii. p. 324
  (1893).

Adult Male.—May be easily distinguished from *P. colchicus*
by having *the lesser and median wing-coverts nearly white*. The
lower back, rump, and upper tail-coverts are more strongly
glossed with purplish-lake in all lights; the black bars down the
middle of the tail-feathers are much narrower; the feathers of
the chest and breast are glossed with purplish-lake and with
much narrower dark purple *margins*, and the middle of the
breast and sides of the belly are purplish-carmine. Total
length, 35 inches; wing, 9·3; tail, 19·5; tarsus, 2·8

Adult Female.—Closely resembles the female of *P. colchicus.*

Range.—South-east of the Caspian Sea, Ashourada Island, and the Peninsula of Potemkin; extending to the east, along the valleys of the Atrak, Sumbar, and Chandir Rivers.

III. THE PRINCE OF WALES' PHEASANT. PHASIANUS PRINCIPALIS.

Phasianus principalis, Sclater, P. Z. S. 1885, p. 322, pl. xxii.; Sharpe, Trans. Linn. Soc. (2) Zool. v. p. 86, pl. vii. (1889); Ogilvie-Grant, Cat. B. Brit. Mus. xxii. p. 325 (1893).

Phasianus komarowii, Bogd. Bull. Acad. St. Petersb. xxx. p. 356 (1886).

Adult Male.—May be easily distinguished by having the white wing-coverts of *P. persicus,* but, unlike that species, the rump is bronze-red, and practically there is *no* purple-lake gloss on the lower back, rump, and upper tail-coverts; the feathers of the chest and breast are *broadly tipped* with purplish-red bronze, and the flank-feathers are broadly tipped with dark purplish-green. Total length, 35·5 inches; wing, 9·4; tail, 21·5; tarsus, 2·7.

Adult Female.—Much paler than the female of *P. colchicus* and *P. persicus,* the ground-colour of the feathers of the mantle being *paler rufous,* and the general colour of the rest of the plumage *pale sandy-buff.* It is extremely similar to the female of *P. chrysomelas,* from the Amu-Darya, described below, having the black spots on the middle of the chest-feathers more strongly marked than in the other allied species.

Range.—North-western Afghanistan and north-east Persia.

Habits.—This extremely handsome species was first discovered in 1884 by the members of the Afghan Delimitation Commission, and Dr. J. E. T. Aitchison, the naturalist attached to the Expedition, prepared some beautiful skins. He informs us that "the specimens of this Pheasant were all got on the banks of the Bala-Morghab, where it occurs in con-

siderable numbers in the tamarisk and grass jungle growing
in the bed of the river. More than four hundred were killed
on the march of thirty miles up this river. It not only wades
through the water in trying to make from one point of vantage
to another, but swims, and seems to be quite at home in these
thickets, where there is always water to the depth of two or
three feet. These swampy localities afford good shelter. In
the mornings and evenings the Pheasants leave it for the more
open and dry country, where they pick up their food. I
believe the same species is found on the Hari-rud river, but
I have seen no specimens from that locality."

IV. THE ZERAFSHAN PHEASANT.   PHASIANUS ZERAFSHANICUS.

Phasianus zerafshanicus, Tarnovski, Field, lxxvii. p. 409
    (1891); Ogilvie-Grant, Cat. B. Brit. Mus. xxii. p. 326
    (1893).
Phasianus klossovskii, Tarnovski, Field, lxxvii. p. 409 (1891).
Phasianus tarnovskii, Seebohm, P. Z. S. 1892, p. 271.

Adult Male.—Closely allied to the male of *P. principalis*, but
the scapulars are *not* margined with dark greenish-purple, and
the breast-feathers have *narrow* heart-shaped purplish margins,
much as in *P. persicus*.

Range.—Zarafshan Valley.

Lieutenant G. Tarnovski writes:—"Mr. Klossovski, who
had resided in Katta-Koorgan for thirteen years, informed me
that Pheasants had made their appearance in the district of
Katta-Koorgan (where we were shooting) about 1883, and that
they had immigrated from the Bokharian dominions, probably
from the Kara-Kul lakes and reeds (the Zarafshan does not
reach the Amu-Darya, but is lost in the sands of Kara-Kul),
whence they were driven forth by the invasion of the Kisil-
Koom sands, which gradually bury the western part of Bok-
hara under their hills.

" At present this Pheasant steadily moves up the Zarafshan.

Thus in the past year, in December, it had been obtained as
high as Goolbah, a village some thirty miles up Samarkand.
It is to be hoped that in a couple of years this species may be
common on the Russian Zarafshan.   On the other hand, *P.*
mongolicus (turned out in 1881-83) has nearly disappeared in
Þagbeet ; in the past year there has been obtained but one
specimen of this species (November, 1890, Lake Doort-Kul),
which tends to prove that it is being crowded out by the
aboriginal species.

   "The Pheasant of the Zarafshan has a mode of life totally
differing from its other Asiatic brethren, owing to the high
state of cultivation of the Zarafshan Valley; it breeds and
nests in reed swamps and marshes bordering on this stream,
and takes its food from the neighbouring fields and gardens.
The best time for obtaining it is just before sunrise, when it
may be found congregated in the reed swamps mentioned
above.   Mr. Klossovski shot in November, 1890, a hen of this
species in male plumage."

   The account given by Lieutenant Tarnovski of the habits of
this bird closely agrees with Dr. Aitchison's account of the
habits of *P. principalis* given above, and there can be no doubt
that the present species is in every way closely allied to it.

   V. SHAW'S PHEASANT.   PHASIANUS SHAWI.

Phasianus shawi, Elliot, P. Z. S. 1870, p. 403 ; id. Monogr.
   Phasian. ii. pl. i. (1872) ; Scully, Stray Feathers, iv. p. 179
   (1876) ; Gould, B. Asia, vii. pl. 35 (1876) ; Sharpe, Second
   Yarkand Miss. Aves, p. 120 (1891) ; Ogilvie-Grant, Cat. B.
   Brit. Mus. xxii. p. 326 (1893).
Phasianus insignis, Elliot, P. Z. S. 1870, p. 404 ; id. Monogr.
   Phasian. ii. pl. iii. (1872).

   Adult Male.—May be distinguished from *P. colchicus* by
having the lesser and median wing-coverts *white* or *whitish-
buff;* the lower back, rump, and upper tail-coverts orange-
bronze with greenish and purplish reflections, the feathers of

the two former with a green spot on each side of the shaft; and the chest- and breast-feathers *edged* with *dark green*. The middle of the breast and sides of the belly are, moreover, *conspicuously dark green*. This species bears the same relationship to *P. chrysomelas*, described below, that *P. persicus* bears to *P. principalis*. Total length, 35·5 inches; wing, 9·6; tail, 19·5; tarsus, 2·7.

Adult Female.—Much paler than the female of *P. colchicus*, the ground-colour of the mantle being *pale rufous-buff*, and the general colour of the rest of the plumage *light buff*. Total length, 23·4 inches; wing, 8·3; tail, 11·2; tarsus, 2·3.

Range.—Valleys of Yarkand and Kashgar, as far east as the Aksu and Khotan rivers.

Habits.—Dr. J. Scully writes:—" This fine Pheasant is a permanent resident in the plains of Eastern Turkestan, frequenting long grass jungle and reeds growing in waste ground. It is said to occur most plentifully in the Dolan jungle, Makit and Maralbashi being mentioned as places where it is particularly numerous. However, it is common enough near Kashgar and Yarkand. I know of two rather good places for this Pheasant, one between Yarkand and Kokrabat, and another near Beshkant. The flight of this bird is rather slow, and it commonly goes over the long grass only for a short distance, and then drops down. When alarmed the male bird utters a harsh, shrill cry.

"These Pheasants are the most untameable birds it is possible to conceive. In confinement they knock their tails to pieces, and wear all the feathers off their heads in insane attempts at escape ; so that a dozen of these birds, after they have been captives for awhile, become the most ragged crew imaginable. Even after being kept in a pheasant-house for months, whenever one approached within a dozen yards of them, they were so alarmed that they would almost knock themselves to pieces, tumble over each other, and fly straight

C 2

upwards, with shrill cries, against the roof of their house.   The
Yarkandis said that even when caught young these birds could
not be tamed.

"The flesh of this Pheasant is, of course, very good eating,
but in my humble opinion does not come up to that of *Tetraogallus tibetanus* (the Tibetan Snow-Cock)."

Young Birds are said to attain full size in about five months.

Nest.—On the ground, in long grass jungle.

Eggs.—From twelve to fifteen in number; like those of *P.
colchicus*, varying in colour from brownish-buff to greyish stone-
colour; mostly a broad oval, slightly compressed towards one
end.   Average measurements, 1·74 by 1·41 inch.

VI. THE TARIM PHEASANT.   PHASIANUS TARIMENSIS.

Phasianus tarimensis, Prjevalsky, Dritte Reise Centr. Asia, p.
95 (1883); Pleske, P. Z. S. 1888, p. 415; Ogilvie-Grant,
Cat. B. Brit. Mus. xxii. p. 327 (1893).

Adult Male.—Closely allied to the male of the last species, *P.
shawi*, but the lesser and median wing-coverts are *yellowish-
brown* instead of whitish; the feathers of the chest and sides of
the breast are fiery bronze-red, glossed with oily green and
purple, and *devoid of marginal bands*, and the margins to the
feathers of the rump and upper tail-coverts are green and buff,
instead of orange-red.   Total length, 30·5 inches; wing, 9·3;
tail, 16·3; tarsus, 2·8.

Adult Female.—Similar to the female of *P. shawi*.

Range.—Extending from Karaschar, along the Tarim Valley,
to Lob-nor.

VII. THE OXUS PHEASANT.   PHASIANUS CHRYSOMELAS.

Phasianus chrysomelas, Severtzov, Bull. Mosc. xlviii. pt. 3, p.
207 (1875); Gould, B. Asia, vii. pl. 36 (1876); Ogilvie-
Grant, Cat. B. Brit. Mus. xxii. p. 327 (1893).

Phasianus dorrandti and *P. oxianus*, Severtzov, J. f. O. 1875, p. 225.

Adult Male.—Easily distinguished from its nearest ally, *P. shawi*, by having a triangular dark green spot at the extremity of each feather of the mantle, back, and rump ; the feathers of the mantle more *widely edged*, and those of the chest, breast, and flanks *very widely tipped with the same colour*. The dark green on the breast and sides of the body much less extensive. Total length, 34 inches ; wing, 9·3 ; tail, 19 ; tarsus, 2·8.

Adult Female.—Resembles the female of *P. shawi*, but, as in the female of *P. principalis*, the black spots on the middle of the chest and breast-feathers are more strongly marked. Total length, 24 inches ; wing, 8·3 ; tail, 12·4 ; tarsus, 2·4.

Range.—Valley of the Oxus or Amu-Darya.

Dr. Severtzov writes :—" My observations on the habits of this Pheasant extend from the month of July to the middle of October. In July they come out from the jungle every morning and evening for the purpose of feeding, and both at sunrise and after sunset their screams may be heard in the bushes ; but day by day towards the end of that month they are seen less and less, and remain more concealed in the thickets. The males are now fast moulting, and the females also, but in a less degree, the latter being then occupied with their chickens. At this time neither males nor females sit on the trees as they do later on, but remain always on the ground, and, from the footprints in the mud, I opine that at this season of the year the moulting Pheasants are actively pursued by the Marsh-Cat (*Felis chaus*). During the night, however, the birds retreat to such thickets as render the noiseless approach of their enemy impossible.

" The birds, as soon as the moult is ended, gather in small flocks, consisting of males, females, and young ; some old males, however, remain single. This association begins with the first days of October, but it is not very strictly kept up.

During the day, numbers of them often disperse among the
bushes, a flock of from ten to fifteen specimens occupying a
space of as many acres, and on being disturbed they fly up one
at a time. They keep more together when feeding in open
places, as, for instance, on the stubble-land. They eat the
seeds of *Eleagnus*, *Halimodendron*, and *Alhagi*. Near the
open spaces covered with the last-named thorny grass they
conceal themselves amongst the tamarisk bushes, in which
they find shelter, but no food. Besides these wild seeds, they
eat in autumn every kind of cultivated corn, particularly *Pani-
cum miliaceum*, as well as peas and lupins. The flocks, though
often dispersed during the day, gather themselves together more
closely at night, which they generally pass in the densest
bushes, as in summer. I have also found them assembling for
the night on the walls of abandoned and deserted farmyards,
which on the Oxus, as well as in Turkestan, are built of clay,
in the form of small fortresses.

" I have never seen a dog bring one of the Pheasants to
perch, as is related of *P. colchicus* in the Caucasus ; and, indeed,
P. chrysomelas is eminently a ground bird, perching only
exceptionally, although commencing to do so at an earlier
season than *P. mongolicus*."

With a broad white ring round the neck.

VIII. THE MONGOLIAN RING-NECKED PHEASANT. PHASIANUS
MONGOLICUS.

Phasianus mongolicus, Brandt. Bull. Acad. St. Pétersb. iii. p.
51 (1844); Gould, B. Asia, vii. pl. 41 (1858); Elliot,
Monogr. Phas. ii. pl. iv. (1872) ; Ogilvie-Grant, Cat. B.
Brit. Mus. xxii. p. 328 (1893).

Adult Male.—Easily distinguished from all the maroon- and
red-rumped species previously described, by having a *broad
white ring* (interrupted in front) round the neck ; otherwise it
most nearly resembles *P. persicus*, but the mantle, chest, and

breast are bronzy orange-red glossed with *purple-carmine* in one light, and green in the other; the rump is dark maroon, strongly glossed with green, shading into purple; the throat is *purplish bronzy-red;* the breast- and flank-feathers are *tipped with very dark green;* and the middle of the breast and sides of the belly are *dark green.* It is, moreover, rather a large bird. Total length, 36·5 inches; wing. 9·6; tail, 22; tarsus, 2·8.

Adult Female.—Much like the female of *P. chrysomelas,* but there is a *black spot* near the extremity of each feather of the *upper mantle* and a *black bar across the middle,* instead of a broad black sub-marginal border. Total length, 26 inches; wing, 8·5; tail, 12·3; tarsus, 2·5.

Range.—From the valley of the Syr-Darya across the basin of Lake Balkash as far east as Lake Saisan and the valley of the Black Irtish, and southwards to the valley of the Ili and Issik-Kul

We can find no notes of importance on the habits of this truly splendid Pheasant.

SUB-SP. *a.* SEVERTZOV'S RING-NECKED PHEASANT. PHASIANUS
SEMITORQUATUS.

Phasianus semitorquatus, Severtz. Ibis, 1875, p. 491; Ogilvie-
Grant, Cat. B. Brit. Mus. xxii. p. 329 (1893).

Adult Male.—Very similar to the male of *P. mongolicus,* but, when skins of the two birds are placed side by side, it will be seen that the mantle, rump, throat, chest, and upper breast of the present bird are glossed with *dull oily green,* instead of purple-carmine, and the white ring is more widely interrupted on the fore-part of the neck.

Adult Female.—Very similar to the female of *P. mongolicus.*

Range.—Dzungaria; in the vicinity of Ebi-nor, Kuldja, Urumtsi, and Gutchen.

Nothing is recorded concerning the habits of this species, but they probably do not differ much from those of the allied forms. Severtzov says that the country between Kuldja and Urumtsi, at the base of the Tian-shan mountains, where this form was first obtained, is a steppe locality with a rivulet and marshes.

B. *General colour of the lower back, rump, and upper tail-coverts greenish or bluish slate-colour, with a rust-coloured patch on each side (except in* P. versicolor). *With a white ring round the neck.*

IX. THE CHINESE RING-NECKED PHEASANT. PHASIANUS TORQUATUS.

Phasianus torquatus, Gmelin, S. N. i. pt. ii. p. 742 (1788); J E. Gray, Ill. Ind. Orn. ii. pl. 41, fig. 1 (1834); Gould, B. Asia, vii. pl. 39 (1856); Sclater and Wolf, Zool. Sket. i. pl. 37 (1861); Elliot, Monogr. Phasian. ii. pl. v. (1872); Prjevalsky, in Rowley's Orn. Misc. ii. p. 385 (1877); Ogilvie-Grant, Cat. B. Brit. Mus. xxii. p. 331 (1893).
Phasianus albotorquatus, Bonnat. Tabl. Encycl. Méth. i. p. 184 (1791).

(*Plate XXII.*)

Adult Male.—The colour of the lower back, &c., mentioned above, serves to distinguish this species at a glance from all those already described. The ground-colour of the mantle and flank-feathers is *bright orange-buff* instead of primrose, as in *P. formosanus* (but it must be added that in some birds from Corea and China this difference is scarcely apparent); the chest- and breast-feathers have only the *narrowest* purple margins, and the whole breast is glossed with purplish-lake, as in *P. persicus.* From the red-rumped species this and the following birds are further distinguished by having the black bars on the basal part of the tail-feathers *much wider.* Total length, 35 inches; wing, 9·2; tail, 20·2; tarsus, 2·7.

Adult Female.—Closely resembles the female of *P. colchicus.* Total length, 24·5 inches; wing, 8·2; tail, 10·5; tarsus, 2·4.

Range.—Extending from the lower Amoor, Mantchuria, Corea, Tsu-sima (Japan), and Eastern Mongolia, through Northern and Eastern China as far south as Canton. Although this species appears to have been found wild by Mr. Holst on the island of Tsu-sima, which is between the Corean coast and Kiusiu, the southern island of Japan, it seems to us probable that it has been introduced there.

On the island of St. Helena, where it was of course introduced, it has long been wild, and numbers are killed annually.

Habits.—The best account of the habits of the Chinese Ring-necked Pheasant is given by Prjevalsky, who writes as follows :—
"We met with the Pheasant north of Gu-bey-key, in Muni-ul, and along the northern bend of the Hoang-ho River. In the former locality they inhabit the wooded districts, usually in the vicinity of brooks, and do not ascend any mountains beyond 6,500 feet above the level of the sea. In the Hoang-ho Valley they keep in the thick groves near Chinese fields and habitations, and drink out of rain-pools or wells, there being only very few brooks and rivulets. The courtship in spring is probably like that of our European Pheasant. The calling of the male reminds one of the voice of a young barn-door fowl, and is followed by a flapping of the wings ; it can be heard at a verst (two-thirds of a mile) distance in clear weather. It usually pairs at the same place, choosing for that purpose the bushes, or some little hill, but never a tree. After each call it remains silent for from five to fifteen minutes, according to the intensity of its excitement and the time of the day. Its calls are loudest and most frequent at sunrise and just before sunset.

"In spring the pairing commences in the beginning of April, and lasts until the end of June, when the males fight vigorously, just like our barn-door fowls, the conqueror pursuing the conquered bird until it is driven off. The hens usually keep close to the cocks, but do not utter any note,

and can be seen in the daytime promenading in company with
them.   At that time these Pheasants are very wild and difficult
to approach ; whilst at all other seasons they are most easy to
shoot with the aid of a dog or by waiting for them at their
drinking-places.

"Immediately the breeding-season is over, the males com
mence moulting, which lasts until October ; and I have seen
some which have lost all their tail-feathers at one time.   In
summer we found in Ordos many families of from six to ten
specimens which were very various in size ; and even as late
as August some young ones were observed which did not
exceed a Ptarmigan in size, and we found that generally
Pheasants breed very late in Mongolia, and that the young
grow slowly.

"Whenever we saw a family of these Pheasants the old birds
were present ; and the male bird seems to look as anxiously
after the young as the hen, and, on the approach of danger,
crows most vigorously, whilst the hen at once takes to wing
and tries to attract the attention of the sportsman and his dog.
The young always endeavour to save themselves by running,
and do not separate from each other until late in the
autumn."

As has been already remarked, since the introduction of
the species into Europe and the British Islands it has inter-
bred freely with *P. colchicus*, and it is now only rarely that one
comes across what appears to be a really pure-bred bird of either
species.   The traces of the Chinese species in nearly pure-bred
P. colchicus are manifest in the partially defined white collar
(sometimes only one or two white feathers) and the green bars
on the feathers of the lower back ; while in nearly pure-bred
P. torquatus, the plumage of the mantle and flanks is darker
than in typical Chinese examples, and the lesser and median
wing-coverts are mixed with sandy-brown.

Crosses between this species and the Japanese *P. versicolor*,
with which it readily interbreeds, are remarkably fine birds,

the male hybrids surpassing in size and beauty the males of either species.

The *nest* and *eggs* are like those of *P. colchicus*.

X. THE SA-TSCHEN RING-NECKED PHEASANT. PHASIANUS SATSCHEUNENSIS.

Phasianus satscheunensis, Prjevalsky, Reisen in Tibet, p. 59 (1884); Ogilvie-Grant, Cat. B. Brit. Mus. xxii. p. 333 (1893).

Adult Male.—This is a very pale form of *P. torquatus*, from which it may be distinguished by having the general colour of the upper-parts much paler, the scapulars and secondary quills being margined with *sandy-brown* instead of Indian-red; the margins to the feathers of the chest and under-parts are wider and *purplish-green*.

Adult Female.—*Very much paler* than the female of *P. torquatus*, the general colour of the upper-parts being pale buff with the black markings much diminished in size; the chin and throat pure white; and the under-parts whitish-buff with faint indications of brown cross-bars on the sides and flanks.

Range.—Sa-tschen, north of the Nan-Shan Mountains.

XI. THE FORMOSAN RING-NECKED PHEASANT. PHASIANUS FORMOSANUS.

Phasianus formosanus, Elliot, P. Z. S. 1870, p. 406; id. Monogr. Phasian. ii. pl. vi. (1872); Ogilvie-Grant, Cat. B. Brit. Mus. xxii. p. 333 (1893).

Adult Male.—Distinguished from *P. torquatus* in having the ground-colour of the mantle and flanks pale primrose; the chest but slightly glossed with pink, and the margins of the feathers more widely edged all round with purplish-green.

Adult Female.—Appears to differ from the *female* of *P. torquatus* in having the barring on the feathers of the chest and breast darker and more distinct.   Only the type specimen, however, has been examined.

Range.—The Island of Formosa.

Habits.—Swinhoe reports the habits of this species as similar to those of its Chinese ally, and says that it affords excellent sport, being particularly numerous.

With no white ring round the neck, or only traces of one.

XII. THE CHINESE RINGLESS PHEASANT.   PHASIANUS DECOLLATUS.

Phasianus decollatus, Swinhoe, P Z. S. 1870, p. 135; Elliot, Monogr. Phasian. ii. pl. vii. (1872); Ogilvie-Grant, Cat. B. Brit. Mus. xxii. p. 331 (1893).

Adult Male.—Very similar to the male of *P. torquatus*, but the white ring surrounding the neck is absent in typical examples, though, where the ranges of the two birds approach one another (for example at the Ichang Gorge), examples of the present species show traces of a white ring, some of the feathers on the hind-neck being banded with white.   It further differs in having the crown of the head *dark green* instead of pale bronze-green, and the margins to the chest-feathers *much broader* and *dark-green* instead of purple.   Total length, 34 inches; wing, 9·2; tail, 18·5; tarsus, 2·8.

Adult Female.—Most like the female of *P. strauchi*, but the black patches, especially those on the scapulars, wing-coverts, and lower back, are *larger and more strongly marked;* and the ground-colour of the mantle is darker chestnut.   Total length, 25 inches; wing, 8·1; tail, 10·5; tarsus, 2·5.

Range.—Western China, extending from Western Yunnan, northwards to Southern Shen-si, eastwards to the Sin-ling Mountains, and southwards to Western Quei-chow.

Habits.—Of all the True Pheasants this is perhaps the rarest in collections, though common enough in many parts of its range. In Western Sze-chuen, Mr. Pratt met with this species on the grassy slopes on the spurs of the mountains up to an elevation of about 9,000 feet. He observed that it avoided the forest regions, always preferring the brushwood, and that in confinement it invariably roosted on the ground. We can find no other notes referring to this species, but no doubt its habits are generally similar to those of *P. torquatus*, of which it is the south-western representative.

XIII. STRAUCH'S PHEASANT. PHASIANUS STRAUCHI.

Phasianus strauchi, Prjevalsky, Mongol. ii. pt. 2, p. 119, pl. xvii. (1876) ; id. in Rowley's Orn. Misc. ii. p. 417 (1877); Ogilvie-Grant, Cat. B. Brit. Mus. xxii. p. 330 (1893).

Adult Male.—Easily distinguished from the males of both *P. elegans* and *P. vlangalii* by having the chest and sides of the breast *fiery orange-red* with narrow, complete, dark purplish-green margins instead of dark green ; from the former it is further distinguished by having the middle of the scapulars *whitish-buff freckled with black* next the shaft, and from the latter by the margins of these feathers being *Indian-red*. Total length, 36·5 inches ; wing, 9·4 ; tail, 23·3 ; tarsus, 2·5.

Adult Female.—Upper-parts much like those of *P. colchicus*, but the feathers of the nape and mantle are indistinctly tipped with dark green, instead of violet and purple ; the underparts are whitish buff barred with black, the bars on the flanks having some green gloss. Total length, 23·5 inches ; wing, 8·1 ; tail, 12·5 ; tarsus, 2·2.

Range.—North-western Kansu.

Habits.—Prjevalsky, who originally discovered and named this very handsome species says :—"The bird inhabits the wooded parts of the Kansu Mountains, up to an absolute height of 10,000 feet. It appears to be most numerous in the Tetunga and Buguk-gol valleys, but higher up these rivers, where woods are scarce, it disappears.

"In voice and habits it does not differ from *P. torquatus* and *P. vlangalii.* The breeding-season commences in April or March, and lasts until the middle of July. The earliest young we obtained on the 23rd of June. The number of young averages from six to ten, and sometimes even twelve; they are always accompanied by both parents; and very often the male bird defends the young even more vigorously than the female."

XIV. VLANGALI'S PHEASANT. PHASIANUS VLANGALII.

Phasianus vlangalii, Prjevalsky, Mongol. ii. pt. 2, p. 116, pl. xvi. (1876); id. in Rowley's Orn. Misc. ii. p. 386 (1877); Ogilvie-Grant, Cat. B. Brit. Mus. xxii. p. 330 (1893).

Adult Male.—May be distinguished from *P. elegans* by having the general colour of the mantle and scapulars *sandy-red*, and the sides and flanks *golden-buff* instead of dull orange-red glossed with purple; from *P. strauchi* it differs in having the colour of the chest *dark green*. Total length, 31·5 inches; wing, 9·5; tail, 17·5; tarsus, 2·6.

Adult Female.—Upper-parts like those of *P. colchicus*, but the predominating colour is *pale buff* and the black markings are much fainter; throat pure white; under-parts whitish-buff with faint brown cross-bars. Total length, 22·5 inches; wing, 8·2; tail, 11; tarsus, 2·4.

Range.—Tsaidam marshes, extending north to the Koko-nor Mountains.

Habits.—Concerning this species, Prjevalsky remarks :—"We found this bird in Tsaidam, where it inhabits the cane-groves and bush-covered localities. In autumn and winter it feeds principally on berries, which it eats while sitting on the branches, and at that time especially is very wild and wary.

It does not differ in voice from *P. torquatus*, and begins to breed very early in spring. We have heard it as early as the 13th of February."

XV. STONE'S PHEASANT. PHASIANUS ELEGANS.

Phasianus elegans, Elliot, Ann. and Mag. N. H. (4) vi. p. 312 (1870); id. Monogr. Phasian. ii. pl. viii. (1872); Ogilvie-Grant, Cat. B. Brit. Mus. xxii. p. 329 (1893).

Phasianus sladeni, Anderson MS.; Elliot, P. Z. S. 1870, pp. 404, 408 ; Anderson, Rep. Zool. W. Yunnan, p. 671 (1878).

Adult Male.—In general appearance this species somewhat resembles a hybrid between *P. colchicus* and *P. versicolor*. Apart from the general bluish slate-colour of the lower back and rump-feathers, which are ornamented with rather wide sub-terminal dark green bands, and the rust-coloured patches on each side of the rump, it has the lesser and median wing-coverts *greenish-grey ;* the chest, upper- and middle-parts of the breast, and the sides of the belly *dark green;* the feathers of the mantle light red with wide dull greenish-bronze margins, and the flank-feathers very similar, but tipped with very dark purplish-green. Total length, **27·5** inches ; wing, 9·1 ; tail, 14·7 ; tarsus, 2·5.

Adult Female.—Differs chiefly from the female of *P. colchicus* in having the throat and fore-part of the neck *white,* and the chest and rest of the under-parts *barred irregularly with black.* It nearly resembles the female of *P. strauchi*, described above. Total length, 21 inches ; wing, 7·9 ; tail, 9·8 ; tarsus, 2·3.

Range.—South-western China, Western Sze-chuen, and West Yunnan.

We can find no record of the habits of this Pheasant ; the two examples sent to the Zoological Gardens by Mr. Stone were obtained in the Yun-ling Mountains, and it was from one of them that Mr. Elliot took his description. Dr. Anderson met with it on the grassy hills in the Momien district of Western Yunnan.

XVI. THE JAPANESE PHEASANT. PHASIANUS VERSICOLOR.

Phasianus versicolor, Vieill. Gal. Ois. ii. p. 23, pl. 205 (1825);
Temm. Pl. Col. v. pls. 6 and 7 [Nos. 486, 493] (1830);
Cassin, Perry's Exp. Jap. ii. p. 223, pl. 1 (1856); Gould,
B. Asia, vii. pl. 40 (1857); Sclater and Wolf, Zool. Sket.
i. pl. 38 (1861); Elliot, Monogr. Phas. ii. pl. ix. (1872);
Ogilvie-Grant, Cat. B. Brit. Mus. xxii. p. 334 (1893).

Adult Male.—Easily distinguished from all other species of
the genus by having the whole of the under-parts *uniform dark
green*. The mantle is *dark green* shot with purple, each feather
being ornamented with *concentric lines of buff*, and there is *no*
rust-red patch on each side of the rump, which is uniform
greenish-slate. In this respect the present species differs from
P. torquatus and all the allied forms with slate-coloured rumps.
Total length, 29 inches; wing, 9·6; tail, 17·5; tarsus, 2·7.

Adult Female.—Much like the female of *P. strauchi*, but the
feathers of the mantle have the centre *almost entirely* black,
with sometimes a thin rufous shaft-stripe and the green tips
are generally conspicuous; the black bars on the breast and
flanks are much more strongly marked. Total length, 24
inches; wing, 8·2; tail, 10·5; tarsus, 2·2.

Range.—The Japanese Islands, except Yezo.

Habits.—Mr. Heine, who met with this beautiful bird on the
hills in the neighbourhood of Simoda, supplies the following
account:—"The walk and ascent had fatigued me somewhat;
I had laid down my gun and game-bag, and was just stooping
to drink from a little spring that trickled from a rock, when,
not ten yards from me, a large Pheasant arose, with loud rustling noise, and before I had recovered my gun, he had disappeared over the brow of a hill. I felt somewhat ashamed
for allowing myself thus to be taken so completely aback; but
noticing the direction in which he had gone, I proceeded more
carefully in pursuit. A small stretch of table-land, which I

soon reached, was covered with short grass and some little clusters of shrubs, with scattered fragments of rocks ; and as I heard a note which I took to be the crowing of a cock Pheasant, at a short distance, I availed myself of the excellent cover, and, crawling cautiously on my hands and knees, I succeeded in approaching him within about fifteen yards. Having the advantage of the wind and a foggy atmosphere, and, being, moreover, concealed by the rocks and some shrubs, I could indulge in quietly observing him and his family. On a small sandy patch was an adult cock and three hens busily engaged in taking their breakfast, which consisted of the berries already mentioned growing hereabouts in abundance. From time to time the lord of this little family stopped in his repast and crowed his shrill war-cry, which was answered by a rival on another hill at some distance. At other moments, again, when the sun broke forth for a short time, all stretched themselves in the golden rays, and, rolling in the sand, shook the morning dew from their fine plumage."

The Japanese Pheasant interbreeds readily with the Chinese Ring-necked Pheasant, the male hybrid being a remarkably fine bird, surpassing in size and beauty either of its parents. This species also crosses freely with *P. colchicus*, the males being truly splendid birds, not unlike *P. elegans* in general colouring, but very much larger. I have been informed that such cross-bred birds are much recommended for turning down in preserves, not only on account of their larger size, but because of their more sedentary habits, for it appears that they seldom stray from the coverts where they have been reared, are less given to running, rise rapidly, and when on the wing fly with greater power.

Females of this species are occasionally met with in male plumage.

Nest and Eggs.—Very similar to those of the Common Pheasant.

I 2

II. *Crown of the head reddish.*

XVII. SŒMMERRING'S COPPER PHEASANT. PHASIANUS SŒMMERRINGI.

Phasianus sœmmerringii, Temm. Pl. Col. v. pls. 8 and 9 [Nos. 487, 488] (1830); Cassin, in Perry's Exped. Japan, ii. p. 225, pl. 2 (1856); Sclater, in Wolf's Zool. Sketches, 2, pl. 32 (1861); Gould, B. Asia, vii. pl. 37 (1867); Elliot, Monogr. Phasian. ii. pl. xii. (1872); Ogilvie-Grant, Cat. B. Brit. Mus. xxii. p. 336 (1893).

Adult Male.—General colour above chestnut or brownish-chestnut, the margins of the feathers of the upper-parts and chest glossed with purplish-carmine, changing to fiery gold; the basal part of the feathers black, most conspicuous on the wing-coverts; breast and rest of the under-parts and tail-feathers chestnut, the long middle pair with wide-set narrow black bars dividing the lighter from the darker chestnut, and the outer pairs widely tipped with black. Total length, 50 inches; wing, 8·8; tail, 37; tarsus, 2·5.

Adult Female.—Crown of the head *blackish*, each feather margined with rufous; general colour of the upper-parts black, mottled with sandy-buff and rufous, the feathers on the mantle with the ground-colour mostly rufous, those of the back and scapulars mostly black, with buff shaft-stripes; chin, throat, and fore-part of the neck pale buff, most of the feathers, except those down the middle, being tipped with black. Chest pale greyish-rufous, spotted with black; rest of the under-parts mostly buff; tail-feathers chestnut, the middle pairs indistinctly mottled with black and buff, the outer pairs tipped with black and white. Total length, 21·0 inches; wing, 8·1; tail, 7·5; tarsus, 2·1.

Range.—The Japanese islands of Hondo and Kiu-siu.

Habits.—Very little has been written about the habits of Sœmmerring's Copper Pheasant, and the only published notes are not very important. Since the year 1865 several birds have bred in the Zoological Gardens, but the young birds have not survived for more than a few days.

Dr. Joseph Wilson gives the following notes on this species :— "During the first part of our stay at Simoda, the cultivated fields afforded no food for the Pheasants. The natives told us they were plentiful in the hills; but no one was willing to undertake to show them, and several rambles through the bushes where these birds were supposed to feed ended in disappointment. Once only I had a glimpse of a brood of young ones, near a hut in the mountains; but they immediately disappeared by running very rapidly. Perhaps one reason of our want of success was to be found in the fact that the wheat was ripe and partially harvested before we left (June 24th), so that during the time of our efforts they were enabled to fill their crops occasionally from the wheat-fields, and lie very close in the hills during the day, without being under the necessity of wandering in search of food.

"The note of one or the other of these species of Pheasants was heard frequently. On the top of a precipitous hill, about a mile south of Simoda, covered by small pines and a very thick growth of shrubbery, a Pheasant (so we were assured by the Japanese) passed the weary hours while his mate was on her nest, and very sensibly solaced himself and her with such music as he was capable of making. It was, however, anything but melodious, and may be represented as a sort of compound of the filing of a saw and the screech of a Peacock. There are two notes only, uttered in quick succession, and represented by the Japanese name of the bird, *Ki-ji;* but the second note is much longer, louder, and more discordant; in fact, has more of the saw-filing character, *Kee-jaeae.* These two notes are uttered; and, if the bird is not disturbed, they are repeated in about five minutes. A good many attempts, perhaps twenty, to become better acquainted with this individual all failed; it seemed impossible to make him fly, though his covert was by no means extensive."

Mr. A. D. Bartlett, the Superintendent of the Zoological
Gardens in London, writing of this species, says :—"Among
the *Phasianidæ* some species are remarkable for their pugnacious and fierce dispositions; not only the males, but
frequently the females, destroy each other. The want of
suffi:ient space and means of escape among bushes, shrubs, or
trees is no doubt the cause of many females being killed when
kept in confinement; and this serious misfortune is unhappily
of no rare occurrence. After the cost and trouble in obtaining
pairs of these beautiful birds, and when they have recovered
from their long confinement on the voyage, their owner is
desirous of reaping a reward by obtaining an abundant supply
of eggs as the birds approach the breeding-season, alas ! he
finds that some disturbance has occurred, the place is filled
with feathers, and the female bird, from which he expected so
much, is found dead or dying, her head scalped, her eyes
picked out, or some other equally serious injury inflicted. I
have found some species more inclined to this cruel practice
than others, the worst, according to my experience, being *P.
sœmmerringii.*"

Eggs.—Pale greenish-white ; rather long ovals ; shell smooth
and fine. Average measurements, 1·8 by 1·35 inch.

VAR. *a.* PHASIANUS SCINTILLANS.

Phasianus (Graphophasianus) scintillans, Gould, Ann. Mag.
  N. H. (3) xvii. p. 150 (1866) ; id. B. Asia, vii. pl. 38
  (1867).

Phasianus sœmmerringii, var. *scintillans*, Elliot, Monogr.
  Phasian. ii. pl. xiii. (1872) ; Ogilvie-Grant, Cat. B. Brit.
  Mus. xxii. p. 337 (1893).

By many authors this form of the Sœmmerring's Pheasant
is regarded as a distinct species; but it can only be recognised
as a well-marked variety, for it not only occurs in the same
islands where *P. sœmmerringii* is found, but every intermediate
stage of plumage between the two forms may be seen.

PLATE XXIII

REEVES'S PHEASANT.

The most typical *male* examples have most of the feathers of the wing-coverts, back, and especially those of the rump, margined on each side with a *white black-edged band*, instead of with fiery gold ; while the lighter parts of the middle tail feathers below the black cross-bars are usually paler, and often strongly dotted with black.

Among the *females* no difference can be observed which is not merely individual or due to age.

III. *Crown of the head white.*

XVIII. REEVES'S PHEASANT. PHASIANUS REEVESII.

Phasianus reevesii, J. E. Gray, in Griff. ed. Cuv. iii. p. 25 (1829); id. Ill. Ind. Zool. i. pl. 39 (1830–32); Sclater in Wolf's Zool. Sketches, 2, pl. 33 (1861) ; Gould, B. Asia, vii. pl. 33 (1869) ; Elliot, Monogr. Phasian. ii. pl. xi. (1872) ; Ogilvie-Grant, Cat. B. Brit. Mus. xxii. p. 337 (1893).

Phasianus veneratus, Temm. Pl. Col. v. pl. 5 [No. 485] (1830).

Syrmaticus reevesii, Wagler, Isis, 1832, p. 1229.

(*Plate XXIII.*)

Adult Male.—Crown *white*, surrounded by a wide black band; chin, throat, and nape *white*, margined below by a black ring which surrounds the neck ; upper-parts *mostly cinnamon*, each feather bordered with black, producing a scale-like appearance; wing-coverts white, broadly margined and centred with black ; chest, sides of breast, and flank-feathers somewhat similar, the two former with chestnut margins, the latter with buff extremities ; rest of under-parts black. Middle pair of tail-feathers enormously elongate, white down the middle, barred with black and chestnut, and brownish-buff on the sides ; outermost pair buff tipped with black. Total length, 6 feet 6 inches ; wing, 10·3 inches; tail, 5 feet ; tarsus, 3·1 inches.

Adult Female.—Crown *reddish-brown ;* rest of head and neck buff except the ear-coverts and a band across the nape, which are mostly blackish-brown; feathers of the upper mantle rufous, tipped with brownish-grey, mottled with black, and each with a *somewhat heart-shaped white spot;* rest of the upper-parts mottled with rufous, buff, and grey, the wing-coverts and scapulars with buff, and the lower back with black, shaft stripes; chest, breast, and sides somewhat like the mantle, but the white spots much less conspicuous; rest of under-parts pale buff; the middle tail-feathers mottled with sandy, buff, and black; outer pairs chestnut mixed with black, and barred and tipped with white.   Total length, 32 inches; wing, 9; tail, 16·6; tarsus, 2·5.

Range.—Mountains of Northern and Western China, extending as far east as Kiu-kiang.

Habits.—The following account of this magnificent Pheasant, the giant of its genus, is given by Mr. E. F. Creagh, in the " Field " of May 13th, 1886, and, though written more from a sportsman's point of view, it gives some idea of the birds' habits in a wild state, and is by far the most interesting note that I have been able to find, very few Europeans having had the good fortune to meet with this Pheasant in its native wilds.

" It was from Ichang, a post at the head waters of the Yangtse, the great river of China, or rather, where that river leaves its gorges, that I started with the stream to a large valley where I knew Reeves's Pheasants had been seen.   It is useless to ask any questions of the country folk, who will always say ' Yes.'   I therefore landed and walked along a wide valley, with high perpendicular mountains of conglomerate on either side, and beetling over small woods of cypress.   The birds live on the berry of this tree, and fly from one wood to another.   They will never show themselves if they can avoid it, and, through their great fleetness when running, steal away from the dogs.   Sometimes, however, when taken by surprise, they rise, and then only by great caution can a single sports-man hope to get them.   Surrounding the small woods with

several guns is the best way to bag them. I think they drive
away the common Pheasant (*P. torquatus*), for I have never
seen the two species together. This may perhaps be due to the
fact of their living on different food. I had with me at the
time a spaniel and a red Irish setter and, as the day was fine
and clear, walked on quietly until I came to what appeared a
good country. The hills here were lower and the wood fairly
dense, but free of undergrowth. A wood-cutter told me he had
seen several Pheasants a few days ago, but could give me no
further information, so, tying up my spaniel, I determined to
work quietly along with the setter. Although it was January,
the day was hot, and I was obliged to divest myself of my coat
as I struggled up the hill. I worked along the lower part without coming on any scent. Suddenly the setter got very busy,
and moved along, showing me that he had some large game.
I followed on, as well as I could, over the broken ground.
False scent, back again; then the dog took a turn up the
almost perpendicular rock. Good gracious! I thought, how
can birds get up there and leave any scent? They had evidently helped themselves with their wings. I was determined
to follow, and brought the setter back to a place where we succeeded in getting on to the upper ledge after a little scrambling.
Having arrived at the top, as I had anticipated, we soon came
on the scent again, and away went the dog, very cautiously
setting every now and again. Just ahead of us now was a
stone wall. I was very much afraid that my game would rise
just as I was getting over, so I made all preparation for a surprise, and at the moment the setter, who had passed the wall,
was at a 'dead set.' I knew there were several birds or some
larger game by the general activity and caution shown by the
dog. I was soon over the wall, ready for anything. I surveyed
my position in a moment. Below me was long grass, on the
edge I had left some thick and high trees, on my right a hill,
also with long, rank grass, but no wood. I moved forward a few
paces, but the dog was there like a marble statue. I was very

badly placed, for I could not see where the game could be. Up
got six Reeves's Pheasants, splendid birds.   I felt certain of
two, but I am sorry to say that I only succeeded in bagging
one, which went rolling down the hill in his last struggles.   I
bounded after him, afraid the dog would mouth his beautiful
plumage.   The bird I had bagged was a cock, measuring five
feet four inches from the bill to end of the tail-feathers.   From
the time I first came on their scent, the distance over which I
had worked must have been a mile."

Reeves's Pheasant has at various times been turned down on
some of the large sporting properties in Great Britain, but it
cannot be considered a success, for the males drive away
the Common and Ring-necked Pheasants and do not interbreed freely with either species.

A pair of these birds was received by Lord Tweedmouth
(then Sir Dudley Marjoribanks) from Pekin in 1870, and turned
out at Guisachen, Inverness-shire, where the breed was successfully maintained for some years, fresh blood being subsequently introduced by the acquisition of four additional
male birds.   Lord Ravensworth makes the following remarks
on the habits of this species as observed by Lord Tweedmouth
in Inverness-shire :—" The Bar-tail is a true Pheasant, well able
to take care of himself in any climate, at any altitude, and is
more easily reared than the common species.   He is very shy
and wild, difficult to approach, and takes to his legs long
before other Pheasants are conscious of any danger.   His
flight is prodigiously rapid and straight, and he will travel
thirty miles on end, which, of course, is an objectionable
practice, except in such extensive forest grounds as the highlands of Scotland present.   These Pheasants travel in troops
of fifteen or twenty, and present a grand and bewildering effect
when they rise in such a company.   Any attempt to walk up to
them in brush covert is utterly hopeless, for they are exceed
ingly vigilant and go straight off like a dart, not more than
six feet from the ground, far out of reach.

"A fight between two old cocks is a beautiful exhibition of activity and spirit. They spring up five or six feet in the air before striking, and s ich is their agility, that the bird assailed hardly ever allows himself to be struck; so much the better for him, for it will be observed that the legs are garnished wiu spurs as long and sharp as those of a game-cock.

"The last peculiarity of this species worth naming is that when they set out on a jaunt, they make for the highest point within range, whereas the Common Pheasant is accustomed to travel downwards along the course of the valleys."

Hybrids between Reeves's and the Golden Pheasant have been bred in confinement, and the males are remarkably handsome birds, having the general plumage reddish-brown.

THE BARRED-BACKED PHEASANTS. GENUS CALOPHASIS.

Calophasis, Elliot, Monogr. Phasian. ii. text to pl. xiii. *bis* (1872).

Type, *C. ellioti* (Swinhoe).

Characters similar to those given for the genus *Phasianus*, but distinguished by having only *sixteen* tail-feathers, and by the males having the lower back and rump transversely barred with black and white.

Only two species are at present known.

I. ELLIOT'S PHEASANT. CALOPHASIS ELLIOTI.

Phasianus ellioti, Swinhoe, P. Z. S. 1872, p. 550; Ogilvie-Grant, Cat. B. Brit. Mus. xxii. p. 335 (1893).
Calophasis ellioti, Elliot, Monogr. Phasian. ii. pl. xiii. *bis* (1872); Gould, B. Asia, vii. pl. 23 (1874).

Adult Male.—Mantle, shoulder-feathers, wing, chest, and breast fiery *bronze-red*, shot with gold; a white band down each side of the mantle; *a band of dark purplish-steel* across the lesser wing-coverts, and *two white bands across the wings* formed by the white ends of the greater wing-coverts and secondaries; throat and fore-neck black; sides of the neck and belly white; *lower back and rump black, barred with white;* tail broadly barred with whitish-grey and chestnut. Total length, 32·5 inches; wing, 8·8; tail, 19; tarsus, 2·8.

Adult Female.—General colour of the plumage pale-drab, barred, mottled, and marked with black on the upper-parts and spotted on the breast; belly mostly white, flanks margined with white; back and sides of neck uniform greyish-drab, *throat and fore-neck black;* outer tail-feathers mostly chestnut with black and white tips. Total length, 20 inches; wing, 8; tail, 7·7; tarsus, 2·5.

Range.—Mountains of South-eastern China.

Habits.—This truly magnificent Pheasant was first discovered by Swinhoe, in the mountains at the back of Ningpo, in the province of Che-Kiang. Subsequently it was met with by Abbé David in Western Fo-kien, where, like the Silver Pheasant, it lives in the wooded mountains, and is far from common, being constantly on the move from place to place, and sometimes remaining away for whole years without revisiting its original habitat.

Eggs. (Laid in confinement.)—Creamy-buff; shell smooth and fine. Average measurements, 1·7 by 1·3 inch.

II. MRS. HUME'S PHEASANT. CALOPHASIS HUMIÆ.

Callophasis humiæ, Hume, Stray Feathers, ix. p. 461 (1880); Godwin-Austen, P. Z. S. 1882, p. 715, pl. 51; Hume, Str. F. xi. p. 302 (1888).

Phasianus humiæ, W. L. Sclater, Ibis, 1891, p. 152; Ogilvie-Grant, Cat. B. Brit. Mus. xxii. p. 335 (1893).

Adult Male.—Differs chiefly from *C. ellioti* in having the neck, upper mantle, and chest glossed with *purplish steel-blue,* like the band across the lesser wing-coverts, but darker; the breast-feathers chestnut, with *steel gloss* and *fiery orange-red margins;*

the belly and flanks *chestnut;* the middle tail-feathers grey, with wide irregular *mixed bars* of chestnut and black, the following pairs barred with black, and the *outermost pairs mostly black,* greyish at the base. Total length, 33·5 inches ; wing, 8·5 ; tail, 20·6 ; tarsus, 2·5.

Adult Female.—Differs chiefly from that of *C. ellioti* in having the throat and fore-neck *devoid of black.*

Range.—Lushai, Manipur, and Chin Hills ; recently obtained near the Ruby Mines in Burma.

Space unfortunately does not permit of my giving in full Mr. A. O. Hume's excellent account [Stray Feathers, xi. pp. 461–467 (1880)] of how he obtained the first known examples of this beautiful Pheasant. The Manipur envoy, who acted as his guide during his travels in the Maharaja's territories, was the proud possessor of a plume of feathers which he was entitled to wear as a mark of rank. Mr. Hume's experienced eye instantly detected among this coveted decoration the tail-feathers of a Pheasant with which he was unacquainted, and after endless enquiries he ascertained that the bird was an inhabitant of the pathless hill jungles of Eastern Lushai and the southern border of Manipur, which had for long been subject to the ravages of the Kamhows, a fierce tribe, who invariably killed everyone they came across. Only one Manipuri of the many questioned had once seen this bird alive in the Jhiri Valley near the Lushai border; to the Maharaja and others it was only known from the tail-feathers which filtered into Manipur through the agency of Kamhow refugees in Manipur. After much trouble, his Manipur guide obtained the services of a party of these Kamhow refugees, who had taken up their abode on the southern borders of the Manipur territory, and in the Eastern Lushai country, and by threatening to shoot the men of this party if they did not return with some of the Pheasants in a short time, entire specimens were at last obtained. Mr. Hume goes on to say, " Sure enough, within the week they returned with one beautiful fresh skin and one perfectly uninjured bird in a cage, both unfortunately males. According to their account, the first day they began trapping they were scented (by their quondam compatriots), their scouts driven in, and they had to fly. This was probably true, because as they were to be paid a large sum per bird, once they were on the ground, they would assuredly not have contented themselves with securing only two. . . .

"The live bird, though a full-grown cock, became perfectly tame in a few days, and was a great favourite in camp. It would eat bread, boiled rice, winged white ants, moths, taking them gingerly out of our hands. At last I thought I really had a prize for the Zoo, something worth sending. Alas, the last day I was in the Eastern Hills, about the middle of the night, the huts in which my servants were, and in which was also my poor Pheasant, suddenly caught fire. . . .

"According to the accounts of my savages, these birds live in dense hill forests at elevations of from 2,500 feet to fully 5,000 feet. They prefer the neighbourhood of streams, and are neither rare nor shy. They extend right through the Kamhow territory into Eastern Lushai and North-west Independent Burmah."

Nest and Eggs.—Unknown.

THE GOLDEN PHEASANTS. GENUS CHRYSOLOPHUS.

Thaumalea, Wagler (*nec* Ruthe, Diptera, 1831), Isis, 1832, p. 1227.

Chrysolophus, J. E. Gray, Ill. Ind. Zool. ii. pl. 41, fig. 2 (1833-4).

Type, *C. pictus* (Linn.).

Tail long and vaulted, composed of *eighteen* feathers, the middle pair being very long, more than four times as long as the short outermost pair.

First primary flight-feather much shorter than the second, which is somewhat shorter than the tenth ; fifth slightly the longest.

GOLDEN PHEASANT.

Male with a full long crest of hairy feathers, and a cape-like development of erectile feathers. Tarsi armed with a pair of short spurs.

Only two species are known.*

I. THE GOLDEN PHEASANT. CHRYSOLOPHUS PICTUS.

Phasianus pictus, Linn. S. N. i. p. 272 (1766); Hayes, Osterl. Menag. p. 5, pls. 5 and 6 (1794).

Thaumalea picta, Wagler, Isis, 1832, p. 1228; Gould, B. Asia, vii. pl. 19 (1866); Elliot, Monogr. Phasian. ii. pl. xv. (1872).

Chrysolophus pictus, J. E. Gray, Ill. Ind. Zool. ii. pl. 41, fig. 2 (1834); Ogilvie-Grant, Cat. B. Brit. Mus. xxii. p. 339 (1893).

(Plate XXIV.)

Adult Male.—*Top of the head, crest,* and rump brilliant *golden-yellow;* square-tipped cape-like feathers covering the back of the neck *brilliant orange,* tipped and banded with black glossed with steel-blue; throat and sides of the head *pale rust-colour;* shoulder-feathers and rest of under-parts *crimson-scarlet,* and middle pair of tail-feathers black, *with rounded spots* of pale brown. Total length about 40 inches; wing, 7·7; tail, 27; tarsus, 2·8.

Adult Female.—Head and mantle brown, barred with black and buff, and mixed with rufous; lower back and rump pale brown, finely mottled with black; throat pale buff; sides of head and rest of under-parts buff, barred with brownish-black except on the middle of the belly. Total length, 24 inches; wing, 7; tail, 14; tarsus, 2·4.

Range.—The mountains of Southern and Western China, extending into Koko-nor.

Hybrids.—The Golden Pheasant crosses freely with the Lady

* In *C. pictus* the parts surrounding the eye are entirely feathered; in *C. amherstiæ* they are naked, but the two species are in all other respects so closely allied that they cannot be separated generically.

Amherst's Pheasant, and the male hybrid is an extremely handsome bird [see Elliot, Monogr. Phasian. ii. pl. xvii. (1872)].

Hybrids have also been produced between this species and the domestic Fowl (Bantam) ; the Common Pheasant (*P. colchicus*) ; and Reeves's Pheasant (*P. reevesi*) ; the last being a large, handsome bird, with almost the entire plumage dull purplish Indian-red.

Eggs. (Laid in confinement.) — Pale creamy-buff; shell rather fine, smooth, and glossy. Average measurements, 1·75 by 1·35 inch.

SUB-SP. *a.* SCHLEGEL'S GOLDEN PHEASANT. CHRYSOLOPHUS OBSCURUS.

Phasianus pictus obscurus, Schl. Ned. Tijdschr. Dierk. ii. p. 152 (1865).

Thaumalea obscura, Elliot, Monogr. Phasian. ii. pl. xvi. (1872).

Chrysolophus obscurus, Ogilvie-Grant, Cat B. Brit. Mus. xxii. p. 341 (1893).

It is extremely doubtful whether this bird has any right to even sub-specific rank. Probably it is merely a domestic variety of the Golden Pheasant, having never been obtained, so far as I am aware, in a wild state.

Adult Male.—Differs from *C. pictus* in having the sides of the head, chin, and throat *brownish-black*, the shoulder-feathers *similarly coloured*, but slightly tinged with red, the outer webs of the flight-feathers devoid of buff margins, and *the middle pair of tail-feathers pale brown, obliquely barred and marked* with black like the second pair.

Adult Female.—Said to differ from the female of *C. pictus* in being generally darker in plumage, especially on the sides of the head and throat. I have never examined a female example of this bird.

II. LADY AMHERST'S PHEASANT. CHRYSOLOPHUS AMHERSTIÆ.

Phasianus amherstiæ, Leadb. Tr. Linn. Soc. xvi. p. 129, pl. 15 (1828).

Thaumalea amherstiæ, Wagler, Isis, 1832, p. 1228; Sclater, List
of Phas. p. 5, pl. 3 (1863); Gould, B. Asia, vii. pl. 20
(1866); Elliot, Monogr. Phasian. ii. p. xx. pl. xiv. (1872).
Chrysolophus amherstiæ, G. R. Gray, List. Gallinæ. Brit. Mus. p.
30 (1867); Ogilvie-Grant, Cat. B. Brit. Mus. xxii. p. 342
(1893).

Adult Male.—Top of the head dark *bronze-green;* long occipital crest *blood-red ;* cape-like feathers covering the back of the
neck *pure white*, margined and barred with black glossed with
steel-blue ; shoulder-feathers, mantle, and chest *dark-green ;*
rump-feathers black, broadly tipped with yellowish-buff; throat
and fore-neck *brownish-black* with some dark greenish gloss ;
rest of under-parts *pure white* barred with black on the flanks;
middle pair of tail-feathers white, with arched black bars on
both webs and wavy black lines across the interspaces. Naked
skin round eye blue. Total length, 50 inches ; wing, 8·2 ;
tail, 36 ; tarsus, 3·1.

Adult Female.—Similar to the female of *C. pictus*, but there is
a naked blue space round the eye as in the male.

Range.—The mountains of Western China and Eastern
Tibet.

Eggs. (Laid in confinement.)—Short stout ovals, pale buff ;
shell smooth, fine, and rather glossy. Average measurements,
1·8 by 1·4 inch.

THE JUNGLE-FOWL. GENUS GALLUS.

Gallus, Linn. Faun. Suecica, p. 61 (1746) ; Temm. Pig. et
Gall. ii. p. 87 (1813).

Type, *Gallus gallus* (Linn.).

Tail composed of *fourteen* or *sixteen* (in *G. varius*) feathers,
laterally compressed and curved downwards, the middle pair (in
the males) being much the longest, about twice as long as the
second pair and nearly four times as long as the outer pair.

First primary flight-feather *considerably shorter* than the tenth ; the fifth slightly the longest.

In the males a high comb extends along the middle of the head from the base of the bill to behind the eyes, the margin being serrated or entire; the sides of the face, chin, and throat are naked, either with two pairs of wattles situated below the ears and on each side of the throat, or with a single wattle (in *G. varius*)* down the middle of the throat ; and the tarsi are armed with long, sharp, curved spurs.

In the females the comb is rudimentary, the wattles absent, the middle tail-feathers are not elongate, and spurs are not developed on the feet.

I. THE RED JUNGLE-FOWL.   GALLUS GALLUS.

Phasianus gallus, Linn. S. N. i. p. 270 (1766).

Tetrao ferrugineus, Gmel. S. N. i. p. 761 (1788).

Gallus bankiva, Temm. Pig. et Gall. ii. p. 87 (1813), iii. p. 654 (1815).

Gallus ferrugineus, Elliot, Monogr. Phasian. ii. pl. 32 (1872) ; Hume and Marshall, Game Birds Ind. i. p. 217, cum tab. (1878); Oates, ed. Hume's Nests and Eggs Ind. B. iii. p. 417 (1890); Tegetmeier, Ibis, 1890, p. 304 [Domestic Breeds].

Gallus gallus, Ogilvie-Grant, Cat. B. Brit. Mus. xxii. p. 344 (1893).

Adult Male.—Long hackles covering the mantle and rump *orange-red or yellowish-orange;*† breast *black*, slightly glossed with green.

In June the hackles and long tail-feathers are moulted both in this and in the following species, and the former are replaced

---

* The Javan Jungle-Fowl (*Gallus varius*) differs, as noted above, from the other species in having *sixteen* tail-feathers and a *single* wattle down the middle of the throat, but, in all other respects, it is a typical *Gallus* and cannot be considered generically distinct.

† There is considerable variation in the colours of the hackles covering the mantle and rump and other parts of the plumage in different specimens, but these differences appear to be merely individual, and are not dependent on locality.

by short black feathers. A second moult takes place in September, and the short feathers of the neck are again replaced by hackles and the long tail-feathers reappear. Total length, 29 inches; wing, 9·5 ; tail, 14 ; tarsus, 3·1.

Adult Female.—Top of the head rust-red shading into orange on the neck and pale yellow on the mantle, each feather with a black stripe down the middle; rest of upper-parts reddish-brown finely mottled with black ; secondary quills mottled with pale reddish-brown towards the edges of the outer webs ; fore-neck chestnut ; rest of under-parts pale light red, browner on the belly and flanks. Total length, 16·5 inches ; wing, 7·5 ; tail, 5·5 ; tarsus, 2·4.

Range.—The jungles of North-eastern and parts of Central India, extending south through the Malay Peninsula to Sumatra and east through Siam to Cochin China. It is also met with in a wild state in Java, Lombock, Celebes, Palawan, the Philippines, Hainan, and other islands, but it seems more than probable that it has been imported at some time or other to all these islands, and that they do not form part of its natural range. It is well known that domestic fowls allowed to escape and run wild in surroundings similar to their original habitat soon revert to the wild type, and become indistinguishable from typical examples of the Indian Red Jungle-Fowl.

All the domestic breeds of poultry are said to have been originally derived from the Red Jungle-Fowl. Some domestic varieties are truly wonderful, not the least so being the Japanese form, in which the hackles covering the tail of the males grow to a length of many feet (as many as fifteen !). Two fine examples of this variety are exhibited in the Central Hall of the Natural History Museum, and are well worth seeing. Equally curious are the black tailless fowls from Holland, a pair of which are shown in the same case.

Habits.—Mr. A. O. Hume, in his "Game Birds of India," writes : "The Red Jungle-Fowl is, as the latter portion of its

E

name imports, a true denizen of the jungle, and most especially
of jungle in the vicinity of scattered cultivation, at or near the
bases of hills, which keep it comparatively well-watered throughout the year.

"It is entirely wanting in the dry, level, alluvial plains, and
semi-deserts of Upper India, and even in better-watered localities is absent from the more richly cultivated tracts, and only
straggles into cultivation which is in the neighbourhood of
jungle. . . . Vertically, this species ranges from sea-level
to 5,000 feet elevation, but, like many other species, it is
generally to be found lower down in the cold season, and
is rarely to be met with above 3,000 feet, except during the
hot season. · . ." Jerdon says : " The Jungle-Fowl is very
partial to bamboo jungle, but is found as well in lofty forests
and in dense thickets. When cultivated land is near their
haunts, they may, during the harvest-season, and after the
grain is cut, be seen, morning and evening, in the fields, often
in straggling parties of ten to twenty. Their crow, which they
give utterance to, morning and evening, all the year round, but
especially at the pairing-season, is quite like that of a Bantam
Cock, but shorter and never prolonged as in our domestic
cocks.

"When detached clumps of jungle or small hills occur in
a jungly district where these fowls abound, very pretty shooting
can be had by driving them by means of dogs and beaters ;
and in travelling through a forest country, many will always be
found near the roads, to which they resort to pick up grain
from the droppings of cattle, &c. Dogs will often put them up,
when they at once fly on to the nearest trees. Young birds, if
kept for a few days, are very excellent eating, having a considerable game flavour."

Colonel Tickell remarks:—"There is no bird more difficult to
approach, or even to see, when in the jungle. The cocks may
be heard of a morning or evening crowing all round, but the
utmost precaution will not, in most cases, enable the sports-

man to creep within shot or sight of the bird. The hen, too,
announces the important fact of having laid an egg with the
same vociferation as in the domestic state, but is silent ere the
stealthiest footstep can approach her hiding-place, and, gliding
with stealthy feet under the dense foliage, is soon far away in
the deep recesses of the jungle. To a stranger it is not a little
curious to hear the familiar sounds of our farmyards issuing
from the depths of the wild forest. . . ."

Mr. A. O. Hume remarks :—" To a certain extent the
Jungle-Fowl *is* omnivorous, and *will* eat not only grass and
young shoots and flower-buds and seeds and grain of all kinds,
but worms and grasshoppers and beetles and small land shells,
but they are preferentially graminivorous, and I have examined
scores which had eaten absolutely nothing but grain.

"In the autumn, after the millet-fields have ripened, they
grow very fat on this grain, and the birds of the year are then
really good eating, but as a rule the birds one kills (be it
confessed with shame, for it *ought* to be a close season), from
March to June, when tiger-shooting in the tarai, when, the day's
sport over, one turns homeward towards the tents, are no whit
better than ordinary village fowls. . . .

"No one specially notices the extreme pugnacity of these
birds in the wild state, or the fact that, where they are numerous,
they select regular fighting-grounds, much like the Ruffs.

"Going through the forest of the Siwáliks in the north-
eastern portion of the Saháranpur district, I chanced one afternoon, late in March, on a tiny open grassy knoll, perhaps ten
yards in diameter and a yard in height. It was covered with
close turf, scratched in many places into holes, and covered over
with Jungle-Fowl feathers to such an extent that I thought
some Bonelli's Eagle, a great enemy of this species, must have
caught and devoured one. Whilst I was looking round, one of
my dogs brought me from somewhere in the jungle round a
freshly-killed Jungle-Cock, in splendid plumage, but with the
base of the skull on one side pierced by what I at once con-

cluded must have been the spur of another cock. I put up for the day at a Bunjara Perow, some two miles distant, and, on speaking to the men, found that they knew the place well, and one of them said that he had repeatedly watched the cocks fighting there, and that he would take me to a tree close by whence I could see it for myself. Long before daylight he guided me to the tree, telling me to climb to the fourth fork, whence, quite concealed, I could look down on the mound. When I got up, it was too dark to see anything, but a glimmer of dawn soon stole into the eastern sky, which I faced; soon after, crowing began all round; then I made out the mound dimly, perhaps thirty yards from the base of the tree and forty from my perch; then it got quite light, and in a few minutes later a Jungle-Cock ran out on to the top of the mound and crowed (for a wild bird) vociferously, clapping his wings and strutting round and round, with his tail raised almost like that of a domestic fowl. . . I learnt so much and no more; there was a rush, a yelp, the Jungle-Cock had vanished, and I found that one of my wretched dogs had got loose, tracked me, and was now careering wildly about the foot of the tree.

"Next day I tried again, but without success. I suppose the birds about had been too much scared by the dog, and I had to leave the place without seeing a fight there; but, putting all the facts together, I had not the smallest doubt that this was the real fighting arena, and that as the Bunjara averred, many of the innumerable cocks in the neighbourhood did systematically do combat there."

Captain Hutton says:—" I have often reared the chicks under a domestic hen, and turned them loose; but, after staying about the house for several days, they always eventually betook themselves to the jungles and disappeared. If kept confined with other fowls, however, they readily interbreed, and the broods will then remain quiet under domestication, and always exhibit, both in plumage and manner, much more of the wild than of the tame stock, preferring at night to roost on

the branches of trees.   Mr. Blyth has remarked that his cross-
bred eggs never produced chicks, but I have never found
any difficulty in this respect.   The crowing of the cock-birds
is very shrill, and like that of the Frizzled Bantams.   In the
wild state it is monogamous."

Mr. Hume remarks again :—" I do not agree with Hutton
that they are always monogamous, because I have constantly
found several hens in company with a single cock, but I have
also repeatedly shot pairs without finding a single other hen in
the neighbourhood; and if you have good dogs (and you can
do nothing in jungle with either these or Pheasants *without*
dogs) you are sure to *see* and *hear*, even if you get no shot at
them, all the birds there are."

Nest.—Generally a shallow hole scraped out of a heap of
dead leaves in any dense thicket, from almost sea-level up to
5,000 feet.   The period of incubation varies from January to
July, according to locality, being earlier farther south.

Eggs.—Usually five or six in number, though as many as
nine eggs are sometimes found, and Major Wardlaw Ramsay
took a nest in Karen-nee containing eleven eggs.

Typically like miniature hen's eggs, but varying much in
size and shape; generally pale yellowish-brown, but occasionally reddish-brown.   Average measurements, 1·78 by 1·36
inch.

II. THE CEYLON JUNGLE-FOWL.   GALLUS LAFAYETTI.

Gallus lafayetii, Lesson, Traité d'Orn. p. 491 (1831).
Gallus stanleyi, J. E. Gray, Ill. Ind. Zool. i. pl. 43, fig. i.
    (1830–32); Hume and Marshall, Game Birds Ind. i.
    pl. (1878).
Gallus lafayetti, Des Murs, Icon. Orn. pl. 18 (1849); Elliot,
    Monogr. Phasian. ii. pl. 33 (1872); Legge, B. Ceylon,
    iii. p. 736, cum tab. (1880); Oates, ed. Hume's Nests and
    Eggs Ind. B. iii. p. 442 (1890); Ogilvie-Grant, Cat. B.
    Brit. Mus. xxii. p. 348 (1893).

Adult Male.—Hackles covering the mantle *golden-orange*, with a black band down the middle, those of the lower back and rump bright orange-red, with a *heart-shaped spot of glossy violet* on the terminal half of each ; chest, breast, and sides, like the lower mantle, orange-red, with a dark maroon stripe down the middle ; belly black, mottled with chestnut.   Total length, 30 inches ; wing, 9 ; tail, 11 ; tarsus, 3.

Adult Female.—Differs chiefly from the female of *G. gallus* in having the secondary quills black and chestnut, irregularly *barred with buff ;* the chest and sides mottled with black and buff, and with whitish centres ; the *breast white, fringed and marked with black.*   Total length, 17 inches ; wing, 7·5 ; tail, 4·9 ; tarsus, 2·4.

Range.—Ceylon.

Habits.—Colonel Vincent Legge, the well-known author of " The Birds of Ceylon," writes as follows :—"The Ceylon Jungle-Fowl inhabits, in abundance, the greater part of the island.   In the low country, it is located in the greatest numbers in the northern, eastern, and south-eastern divisions, which, covered with jungle and possessed of a dry climate, are specially suitable to the habits of the birds. . . .

" The cock-birds are, as is the case with other species, most pugnacious, and pass their time in the mornings and evenings in giving out their well-known challenge-call, ' *Cluk George Joyce*,' accompanied with the usual galline flap of the wings. By using a pocket-handkerchief doubled up into a ball, placed in the palm of the hand, and struck with the other, this sound can be fairly imitated; and if the sportsman be out of sight, well concealed in a hollow in the ground, or behind a huge log or stump, the cocks can be enticed near enough to be shot ; they are so shy, however, that if the least sound be made other than this flapping, they turn round and disappear at once into the thicket.   The natives produce the required sound by striking the thigh with the open hand, slightly

JAVAN JUNGLE-FOWL.

curved; and both Cingalese and Tamils shoot the Jungle-Fowl for the market by thus decoying them.

"While challenging each other, the males often wander close to paths and tracks through the jungle, and still keep up their call, although people may be passing, and laughing and shouting going on; but directly you strike off the road to stalk them, the sound of footsteps puts an end to the *George Joyce*, and the pugnacious bird may be heard rapidly beating a retreat over the fallen leaves.

"At night they roost on trees, but do not choose very high branches, generally seating themselves across a moderately elevated horizontal limb, and, when going to rest, they utter a clucking note very different to the ordinary call.

"The hens are seldom seen near the cocks, and are very shy; they may be sometimes surprised in the early morning scratching by the sides of the roads with their young brood, but on the whole are much more seldom observed than the other sex."

Nest.—Similar and similarly situated to that of the Red Jungle-Fowl. The period of incubation ranges from January to August.

Eggs.—Two to four in number. Like those of the Red Jungle-Fowl, but minutely speckled all over with brownish-red, often with a few somewhat larger spots about the more obtuse end. Average measurements, 1·71 by 1·30 inch.

III. THE GREY JUNGLE-FOWL. GALLUS SONNERATI.

Gallus sonnerati, Temm. Pig. et Gall. ii. p. 246 (1813); iii. p. 659 (1815); Temm. Pl. Col. v. pls. 1 and 2 [Nos. 232, 233] (1823); Elliot, Monogr. Phasian. ii. pl. 34 (1872); Gould, B. Asia, vii. pl. 56 (1877), Hume and Marshall, Game Birds Ind. i. p. 231, pl. (1878); Oates, ed. Hume's Nests and Eggs Ind. B. iii. p. 420 (1890); Ogilvie-Grant, Cat. B. Brit. Mus. xxii. p. 350 (1893).

Phasianus indicus, Leach, Zool. Misc. ii. p. 6, pl. 61 (1815).

(*Plate XXV.*)

Adult Male.—Hackles covering the back of the neck and mantle black, fringed with grey, and with a *yellowish spot like sealing wax* at the extremity; a patch of wing-coverts similarly marked, but the wax-like spots longer and orange-chestnut; hackles on sides of rump similarly ornamented, but fringed with reddish-orange; lower back and under-parts black, glossed with purple, and edged with grey, and with narrow white shaft stripes. Total length, 28 inches; wing, 9; tail, 16·5; tarsus, 3·1.

Adult Female.—Most like the female of *G. lafayetti*, the breast-feathers being *white*, fringed and marked with black, but the outer webs of the secondary quills *finely mottled* with sandy-brown and black, and the breast-feathers *devoid* of the thick black cross-bars. Total length, 16·5 inches; wing, 8·0; tail, 5·2; tarsus, 2·7.

Range.—Western, Southern, and Central India, bounded on the north-east by the valley of the Godaveri, and on the north-west by the Aravalli Hills.

Habits.—Like their northern congeners, these are eminently birds of jungly and hilly or broken ground, and are not to be found at any distance from these in level, thoroughly cultivated, tracts; but throughout all the hilly tracts within the limits indicated, the entire range of the Western Ghâts, the Sátpuras, and all their southern ramifications, the Nilgiris, Pulneys, Anamallis, Shervaroys, and the like, they occur, and, where not persecuted, in great abundance, from near sea-level to at least 5,500 feet elevation. Indeed, individuals may be met with up to fully 7,000 feet, as on the higher slopes of Doda-betta. (*Hume.*)

The late Mr. W. R. Davison writes:—" Unlike the Red Jungle-Fowl, this species is not gregarious, and though occasionally one meets with small coveys, these always consist of only one or two adults, the rest being more or less immature. As a rule, they are met with singly or in pairs. The crow of the male is very peculiar, and might be syllabled '*Kuck-Kaya-Kya-Kuck*,' ending with a low double syllable, like '*Kyu-kun, Kyukun*,' repeated slowly, and very softly, so that it cannot be heard except when one is very close to the bird. Only the males crow, and that normally only in the mornings and evenings, though occasionally they crow at intervals during the day when the weather is cloudy. The crow is very easily imitated, and with a little practice the wild birds may be readily induced to answer.

"They do not, however, crow the whole year through, but only from October to May, when they are in full plumage.

"When flushed by a dog in the jungle, they flutter up into some tree above with a peculiar cackle, a '*Kuck-kuck-kuck*,' which, however, they only continue till they alight.

"They come into the open in the mornings and evenings, retiring to cover during the heat of the day, unless the weather is cloudy, when they may be met with in the open throughout the day.

"Though found in evergreen forests, they seem to prefer moderately thin and bamboo jungle.

"Ordinarily, as already remarked, they are found scattered; but when a tract of bamboo comes into seed, or any other particular food is locally abundant, they collect there in vast numbers, dispersing again as soon as the food is consumed. . . .

"In some ways they are not very shy; by taking an early stroll, even without a dog, along some quiet road by which cattle and grain pass, several can always be obtained, but when they have been at all disturbed and shot at, they become very wary, and even with a dog, before which they ordinarily perch at once, they are very difficult to secure. In such cases, they run till they think they are out of shot, and then rise, and instead of perching, take a long flight, often of many hundred yards, and when they do alight, commence running again. . . .

"The best time to shoot the Jungle-Cock is from October

to the end of May, as then his hackles are in the best condition.

"In June the moult begins, and the male gradually drops his hackles and long tail-feathers, the hackles being replaced by short feathers as in the female; during the rains the male is a poor mean-looking object, not in the least like his handsome self in the cold weather, and, fully conscious of this fact, he religiously holds his tongue during this period.

"In September, a second moult takes place; the short feathers of the neck are again replaced by the hackles, the long tail-feathers reappear, and by October the moult is complete, and our Southern Chanticleer as noisy as ever.

"The male usually carries his tail low, and when running, he does so with the tail lowered still more, the neck outstretched, and the whole body in a crouching position, as in the Pheasants.

"I do not know for certain whether the species is polygamous or monogamous, but from what I have observed I should think the latter; for although the male does not, I believe, assist in incubation, yet when the chicks are hatched, he is often to be found in company with his mate and little ones.

"These birds are, I believe, quite untamable, even when reared from the egg, and though in the latter case they may not be so wild as those captured in maturity, they never take kindly to domestic life, and avail themselves of the first opportunity for escaping. It is needless to say that they cannot easily be induced to breed in captivity. I have known the experiment tried time after time unsuccessfully."

Nest.—Like that of the Red Jungle-Fowl. The period of incubation appears to vary much in different localities, eggs having been found from October to June.

Eggs.—Vary much in shape and colour, the extreme types being a long oval with fine shell, and a broad oval conspicuously

pitted all over with pores. The colour varies from pale creamy-white to rich brownish-buff. Every intermediate form between these two types can be found. Average measurements, 1·84 by 1·38 inch.

IV. THE JAVAN JUNGLE-FOWL. GALLUS VARIUS.

Phasianus varius, Shaw and Nodder, Nat. Misc. x. pl. 353.
Gallus varius, Elliot, Monogr. Phasian. ii. pl. 35 (1872); Ogilvie-Grant, Cat. B. Brit. Mus. xxii. p. 352 (1893).
Gallus furcatus, Temm. Pig. et Gall. ii. p. 261 (1813); iii. p. 662 (1815); id., Pl. Col. v. pl. iv. [No. 483] (1829); J. E. Gray, Ill. Ind. Zool. i. pl. 43, fig. 2 (1830–32).
Gallus javanicus, Horsf. Tr. Linn. Soc. xiii. p. 185 (1822).

(*Plate XXVI.*)

Adult Male.—Easily recognised from the three species previously mentioned by having the *margin* of the comb *entire,* and a *single* wattle down the middle of the throat. Back of the neck and upper mantle covered with *short square-tipped feathers,* purplish-blue, edged with greenish-bronze; lower mantle golden-green, shot with blue, shading into violet-bronze towards the tips and all the feathers margined with black; hackles of the lower back long and black, narrowly margined with yellow; lesser and median wing-coverts elongate, orange-red, with a broad black stripe down the middle; under-parts *black.* Total length, 28 inches; wing, 9; tail, 11·5; tarsus, 2·9.

Adult Female.—Neck and upper mantle sandy-brown, the feathers of the latter with sub-marginal blackish bands and dusky centres; rest of upper-parts *black,* slightly glossed with green, *and irregularly barred and margined with buff;* throat white; *under-parts buff,* with dark margins to the chest-feathers and mottlings on the flanks. Total length, 15·3 inches; wing, 7·7; tail, 4·5; tarsus, 2·3.

Range.—Java, Lombock, and Flores. It is also said to have been obtained wild near Singapore, but no doubt had been imported.

Among the doubtfully distinct species of Jungle-Fowl is a bird which has been named *Gallus æneus* [Temm. Pl. Col. v. pl. 3 [No. 374] (1825)]. The type, which is preserved in the Paris Museum, was obtained in Sumatra, and it is just possible that it may prove to be a distinct species, but at present it is believed by most ornithologists to be merely a hybrid between the Domestic Fowl and the Javan Jungle-Fowl.

Another hybrid bird about which there can be very little doubt has been named and figured by G. R. Gray under the name *Gallus temminckii* (P. Z. S. 1849, p. 62, pl. vii.). The type of this bird is in the British Museum, and is said to have come from Batavia. In the opinion of the writer, it is a hybrid between the Red and Javan Jungle-Fowls, and clearly exhibits the characters of both parents. The comb is toothed, and in addition to the large median throat wattle, there is a small lateral pair. *Gallus violaceus*, Kelsall (J. As. Soc. Str. Br. No. xxiv. p. 167 (1891) and No. xxv. p. 173) is also in all probability a hybrid between the above species. The type, supposed to have come from Borneo, was living in the Botanic Gardens at Singapore. The bird from the Sulu Islands has been named *Gallus stramineicollis* by Dr. Sharpe (P. Z. S. 1879, p. 317), but it is probably only the offspring of a domestic variety run wild.

THE PEACOCK-PHEASANTS.   GENUS POLYPLECTRON.

Polyplectron, Temm. Pig. et Gall. ii. p. 363 (1813).

Type, *P. chinquis* (Müll.).

Tail composed of *twenty to twenty-four* feathers, large, full, and rounded, the middle pair being about twice as long as the outer pair, and each ornamented with *one or a pair of metallic ocelli,* which are rudimentary or absent in the females of some species.

PLATE XXVII.

Wyman & Sons Limited

GREY PEACOCK-PHEASANT.

First primary flight-feather much shorter than the second, which *is shorter than* the tenth ; sixth rather the longest.

Sides of the face *naked*, or very nearly so.

The male is generally armed with two, and sometimes three spurs on each leg. Sexes different in plumage.

I. THE GREY PEACOCK-PHEASANT.   POLYPLECTRON CHINQUIS.

Pavo chinquis, Müll. Linn. S. N. Suppl. p. 121 (1776).

Pavo tibetanus, Gmel. S. N. i. pt. ii. p. 731 (1788).

Polyplectron c'hinquis, Temm. Pig. et Gall. ii. p. 363 (1813), iii. p. 675 (1815) [part] ; id. Pl. Col. v. pl. 17 [No. 539] (1831) ; Gould, B. Asia, vii. pl. 50 (1871) ; Sclater, P. Z. S. 1879, pl. viii. fig. 2 ; Ogilvie-Grant, Cat. B. Brit. Mus. xxii. p. 354 (1893).

Polyplectron tibetanum, Elliot, Monogr. Phasian. i. pl. 6 (1872) ; Hume and Marshall, Game Birds Ind. i. p. 105, pl. (1878).

Polyplectron atelospilum, P. cyclospilum, P. enicospilum, G. R. Gray, List Gall. Brit. Mus. pp. 23, 24 (1867).

Polyplectron helenæ, Oates, Ibis, 1883, p. 136, pl. v. (immature).

(*Plate XXVII.*)

Adult Male.—Upper parts mostly *brown,* dotted with dirty *white ;* near the tip of most of the feathers a large round eye of metallic green and violet, changing to purple and blue, and edged with successive rings of black, brown, and dirty white ; longer upper tail-coverts and tail-feathers, with divided *pairs* of large oval ocelli, *one on each side of the shaft,* entirely green in one light and purple in the other; *throat white;* rest of under-parts *brown with irregular dotted and mottled bars of dirty white ;* naked skin on side of face pale fleshy-yellow. Total length, 25·5 inches ; wing, 8·4 ; tail, 13·2 ; tarsus, 3.

Adult Female.—Upper-parts *brown,* mottled with *pale brown.* Differs from the male chiefly in having the ocelli on the upper-parts represented by black spots, with some violet and purple gloss ; the ocelli on the longer upper tail-coverts *absent,* and only obscurely represented on *both webs* of the *outer pairs of tail-feathers,* those on the middle pairs being rudimentary, and represented by undefined black spots. Total length, 19 inches ; wing, 7·1 ; tail, 8 ; tarsus, 2·4.

Range.—Indo-Chinese countries, extending in the north-west to Sikhim, eastwards to the Laos country, and south-wards through Tenasserim as far as Mergui.

Mr. A. O. Hume writes :—" This species occurs at very varying elevations. I have received it from places in Cachar and Sylhet, and from the base of Nwalabo in Tenasserim, from localities little above sea-level, while, again, Davison obtained it almost at the summit of Mooleyit, at quite 6,000 feet elevation. But though it occurs right down on the plains, it is so far a hill-Pheasant that it chiefly affects hills and their immediate neighbourhood, and is never found in any considerable numbers at any great distance from these."

Mr. R. A. Clark gives the following information :—" The Peacock-Pheasant is very common in North-eastern Cachar, where it is found in dense bamboo jungle, on the sides of ravines, and on the tops of the low ranges of hills where-ever there are *Jámun* trees, as well as on the banks of the river ' Barak,' wherever it is well-wooded. On the rocky faces of the ' Barak ' banks there is a tree which, during the rainy season, is partially submerged, but in the cold weather bears a fruit with seeds like those of a ' chilli.' On these the birds feed greedily in the early morning and towards sunset. Insects and worms, with this fruit, form their chief food, but I have on one occasion found small land shells and pebbles in the stomach of an adult male.

" These birds may be heard in the early morning and at sunset calling, and then the male is generally to be found perched on some branch only a few feet off the ground. The call is *Ha-ha-ha-ha,* something like a laugh, and can be heard from a good distance ; the female's note I have never heard.

" From November to April these birds are found all over

the well-wooded parts of the district ; and during the rainy season they retire to the dense forests and bamboo jungle to breed, and at this season the call is never heard.

"I have shot dozens of this bird, some of which had two and three spurs, but in no case did I ever see more than four on one leg, and one peculiarity is that they hardly ever have the same number of spurs on each leg. The Kookies have an idea that an additional spur grows every year ; but, during the five years' experience I had of them, I never saw more than the number mentioned above. The females have a corn on each leg where the spur is in the male.

"These birds go about in pairs generally, but on one occasion, in December, while riding through a forest pathway, I came across a party of four, one male and three females, the latter easily distinguishable by their smaller size and duller colours.

"As a rule, these Pheasants are very shy, and terrible runners and skulks, and without a good dog it is impossible to secure a winged bird. They are delicious eating. . . ."

Writing from North-east Cachar, Mr. Inglis remarks :— "The Kookies snare numbers of the *Polyplectron* on their '*jhooms*,' or cultivation clearings, inside the forests. The snare consists generally of a sapling, or branch of a tree, bent towards the ground ; one end of a piece of string is fastened to the sapling, and on the other end is a noose ; the latter is spread round a small hole in the earth ; the trap itself is a simple contrivance of a few split pieces of bamboo ; the bait is a small red berry of which the bird is very fond ; the berry is firmly attached to the trap, and the bird pecking at the berry releases the catch, the sapling flies up, and the bird is noosed by the neck or feet, or sometimes both."

We are told that when the young of this species were first hatched in the Zoological Gardens, a Bantam Hen was employed as a foster-mother, and that the chicks *would* follow close behind her, never coming in front to take food,

so that, in scratching the ground, she frequently struck them
with her feet. The reason for the young keeping in her rear
was not understood until, on a subsequent occasion, two
chicks were reared by a hen *P. chinquis*, when it was observed that they always kept in the same manner close behind
the mother, who held her tail widely spread, thus completely
covering them, and there they continually remained out of
sight, only running forward when called by the hen to pick up
some food she had found, and then immediately retreating to
their shelter.

Nest.—A circular depression in the ground made of leaves
and twigs, slightly lined with a few of the birds' own feathers,
placed amongst grass among jungle (*R. A. Clark*).

Eggs.—Like those of the Golden Pheasant. Average
measurements, 2 by 1·44 inches (*R. A. Clark*).

II. GERMAIN'S PEACOCK-PHEASANT.   POLYPLECTRON GERMAINI.

Polyplectron germaini, Elliot, Ibis, 1866, p. 56 ; id. Monogr
Phasian. i. pl. 8 (1872); Ogilvie-Grant, Cat. B. Brit. Mus.
xxii. p. 357 (1893).

Adult Male.—Like the male of *P. chinquis*, but the whitish-
brown spots on the upper-parts are *much smaller and closer
together ;* the ocelli on the tail-feathers dark green with bluish-
violet reflections ; *no* white on the throat. Naked skin round
the eye *crimson.* Total length, 20 inches; wing, 7·5 ; tail,
10·5 ; tarsus, 2·5.

Adult Female.—Differs chiefly from the female of *P. chinquis*
in having the ocelli on the mantle and upper tail-coverts,
though rudimentary, comparatively well-marked and glossed
with dark green, changing to purple ; those on the scapulars
and secondary quills bordered above with narrow Λ-shaped

NAPOLEON'S PEACOCK-PHEASANT.

black and buff bands. Total length, 18·6 inches; wing, 7; tail, 8·5; tarsus, 2·3.

Range.—Cochin China.

III. THE MALAYAN PEACOCK-PHEASANT. POLYPLECTRON BICALCARATUM.

Pavo bicalcaratus, Linn. S. N. i. p. 268 (1766).

Polyplectron bicalcaratum, Sclater, List of Phas. p. 12, pl. 12 (1863); Gould, B. Asia, vii. pl. 51 (1870); Elliot, Monogr. Phasian. i. pl. 7 (1872); Hume and Marshall, Game Birds Ind. i. p. 113, pl. (1878); Ogilvie-Grant, Cat. B. Brit. Mus. xxii. p. 357 (1893).

Polyplectron hardwickii, J. E. Gray, Ill. Ind. Zool. i. pl. 37, ii. pl. 42, fig. 1, and *P. lineatum*, id. *op. cit.* i. pl. 38 (1830–32).

Adult Male.—A short crest glossed with purplish-green; bars on neck-feathers glossed with purple; ground-colour of the upper-parts *buff*, thickly dotted and mottled with *black;* the ocelli on the wings and tail-feathers entirely rich green in one light, and purplish or blue in another; only the middle pair of tail-feathers, and longer upper tail-coverts with ocelli on each side of the shaft, and these are *confluent, and not divided from one another* by a pale band; on all the outer tail-feathers, *only one ocellus on the outer web;* naked skin round the eye red. Total length, 22 inches; wing, 8·2; tail, 10; tarsus, 2·8.

Adult Female.—Distinguished from the female of *P. chinquis* by having the ground-colour of the back, wings, and rump brownish buff, mottled *with black;* feathers of the *mantle and wings with a black blotch* near the tip; *longer upper tail-coverts and middle pair of tail-feathers with paired ocelli;* outer tail feathers with *no* ocelli on the inner webs. Total length, 18 inches; wing, 7·1; tail, 6·5; tarsus, 2·3.

Range.—From the Malay Peninsula and Sumatra to the South Tenasserim.

IV. THE BORNEAN PEACOCK-PHEASANT.   POLYPLECTRON
SCHLEIERMACHERI.

Polyplectron schleiermacheri, Brügg. Abhandl. Ver. Brem. v.
p. 461, pl. ix. (1877); Ogilvie-Grant, Cat. B. Brit. Mus.
xxii. p. 359 (1893).

Adult Male.—Most nearly allied to the male of *P. bicalcaratum*, but the crest much shorter, the frill of feathers round
the hind-neck more developed, and the violet margins much
wider; the metallic ocelli on the back and wings bluish-
green; feathers on the sides of the neck and chest *with heart-
shaped metallic spots* of golden-green, changing to purplish-
blue; rest of the under-parts *black*, with some buff speckling;
middle of neck, breast, and belly, *white;* the paired ocelli on
the upper tail-coverts and middle tail-feathers touching one
another, but not confluent. Total length, 20 inches; wing,
7·8; tail, 8·0; tarsus, 2·6.

Adult Female.—Differs from the female of *P. bicalcaratum*
in having *no* ocelli on the *longer upper tail-coverts*, so that the
tail shows only *a single* series; the under-parts washed with
black. Total length, 14 inches; wing, 6·8; tail, 5·6; tarsus,
2·3.

Range.—Borneo.

This species was first discovered by Herr Schleiermacher in
the interior of Dutch Borneo, and Mr. A. H. Everett has
since met with it at Paitan in the northern part of the island.

V. NAPOLEON'S PEACOCK-PHEASANT.   POLYPLECTRON
NAPOLEONIS.

Polyplectron napoleonis, Less. Traité d'Orn. pp. 487, 650
(1831); Ogilvie-Grant, Cat. B. Brit. Mus. xxii. p. 361
(1893); Bourns and Worcester, Occ. Pap. Minnesota
Acad. i. no. i. p. 43 (1894).
Polyplectron emphanum, Temm. Pl. Col. v. pl. 18 [No. 540]
(1831); Elliot, Monogr. Phasian. i. pl. 9 (1872).

Polyplectron nehrkornæ, Blasius, Mitth. Orn. Ver. Wien. 1891, p. i.; Ogilvie-Grant, Cat. B. Brit. Mus. xxii. p. 36c (1893).

(Plate XXVIII.)

Adult Male.—Crown and *elongate crest* dark green, shot with purplish-blue; back of the neck, mantle, and wings, black, broadly barred with greenish-blue, changing to dark blue and deep violet and fringed with golden-green; lower back and tail black, thickly spotted with rufous-buff, the longer upper tail-coverts, and tail-feathers with a pair of bluish-green ocelli changing to deep violet, each framed in a black and grey ring; the latter absent on the inner webs of the outer pairs of tail-feathers; the broad superciliary stripes* confluent on the nape, and a *triangular patch on each side of the head, pure white;* throat and under-parts *black;* naked skin round eye crimson. Total length, 19 inches; wing, 7·3; tail, 8·7; tarsus, 2·5.

Adult Female.—A *well-developed crest* brownish-black; upperparts pale rufous-brown, finely mottled with black; feathers of the mantle and wing-coverts with an ill-defined band of rufous-buff near the extremity; longer tail-coverts *without* ocelli; ocelli on the tail-feathers as in the male, but much smaller and more widely edged with black only; feathers on sides of face

---

* Two species have generally been recognised, distinguished by the presence or absence of a white eyebrow. On this subject Messrs. Bourns and Worcester observe :—" While in Palawan we were so fortunate as to secure a series of eleven fully adult males of the *Polyplectron* inhabiting that island. Of these, *two have not the slightest trace* of superciliary stripes, while a third has only four small white feathers on one side. In each of the above there are a few white feathers on the nape. Three of our specimens perfectly agree with the description of typical *P. nehrkornæ*. Three specimens have *broad* superciliary stripes, *almost* confluent on the nape, and in one bird the superciliary stripes which beg'n between the eye and the nostril are very broad, widening steadily towards the nape, *where they are fully confluent*. An examination of young birds, of which we have a good series, shows that the width and extent of the superciliary lines is independent of age. We therefore feel perfectly satisfied that *P. napoleonis* and *P. nehrkornæ* are identical, since the width of the white superciliary stripes is an uncertain quantity, subject to wide individual variation, and it may even be absent."

and throat white; under-parts reddish-brown, faintly mottled with black and with pale shafts; orbital skin black. Total length, 14·5 inches; wing, 6·8; tail, 5·5; tarsus, 2·2.

Range.—Island of Palawan.

Habits.—Mr. J. Whitehead, who is one of the few Europeans who have trapped this bird in its native wilds, says:—"This splendid little Pheasant is scarce and local, all my specimens having been collected in one forest; and although my men set hundreds of snares in other forests we never met with another during our three months' stay. One female was eaten by a wild cat in one of the traps, and I rather suspect that this little tiger destroys numbers of this beautiful bird.

"This species, like the Argus Pheasant, has its 'showing off' arena, a neatly swept patch some three or four feet in diameter; the chosen spot is generally in some unfrequented part of the forest. I often noticed that this ring had a small hump of earth in the middle, where no doubt the male birds show off their splendid plumage and perhaps do battle. Their battles, if they have any, must be very short and decisive, as the double spurs of the cock would be sufficient to cut his adversary into bits.

"I am inclined to think that the birds pair, and are not polygamous, as we collected three pairs; but that was not during the nesting-season, which is probably in the months of December and January."

Messrs. Bourns and Worcester add:—"*P. napoleonis* is extremely shy. All our specimens were snared, though Mr. Bourns caught a glimpse of a female on one occasion."

THE BRONZE-TAILED PEACOCK-PHEASANTS. GENUS CHALCURUS.

Chalcurus, Bonap. Ann. Sci. Nat. (4) i. p. 146 (1854).

Type, *C. chalcurus* (Less.).

Tail elongate and wedge-shaped, as in *Phasianus*, and composed of *sixteen* feathers, the middle pair nearly three times as long as the outer pair, and all *partially glossed with metallic purple, but without metallic ocelli.*

First primary flight-feather much shorter than the second, which is *shorter than* the tenth ; sixth slightly the longest.

Sides of the face *covered with small feathers.*

Male armed with two or more pairs of spurs. Plumage of sexes alike.

Only one species is known.

I. THE BRONZE-TAILED PEACOCK-PHEASANT. CHALCURUS CHALCURUS.

Polyplectrum chalcurum, Less. Traité d'Orn. p. 487 (1831).
Polyplectron chalcurum, Temm. Pl. Col. v. pl. 19 [No. 519] (1831) ; Elliot, Monogr. Phasian. i. pl. 10 (1872).
Chalcurus inocellatus (Less.), Bonap. C. R. xlii. p. 878 (1856).
Chalcurus chalcurus, Ogilvie Grant, Cat. B. Brit. Mus. xxii. p. 361 (1893).

Adult Male.—General colour *brown ;* upper-parts barred and fringed with dull rufous ; throat- and neck feathers with white shafts ; middle tail-feathers black, irregularly barred with rufous and shading into *metallic purplish-blue* towards the extremities ; outer pairs similar but with the *greater part of the outer and much of the inner webs purplish-blue,* changing into violet. Total length, 18 inches ; wing, 6·6 , tail, 9·2 ; tarsus, 2·4.

Adult Female.—Like the male, but smaller and devoid of spurs. Total length, 15·7 inches ; wing, 6·2 ; tail, 7·4 ; tarsus, 2·2.

Range.—Sumatra.

Nothing is known of the habits of this rare Pheasant. Dr Büttikofer tells us that the native name is " Karo-Karo."

THE ARGUS PHEASANTS. GENUS ARGUSIANUS.

Argus, Temm. (*nec* Boh. Mollusca, 1761 ; *nec* Scop. Lepidoptera, 1777), Pig. et Gall. ii. p. 410 (1813).

Argusianus, Rafinesque, Analyse, p. 219 (1815).

Type, *A. argus* (Linn.).

Tail composed of *twelve feathers*; the middle pair enormously elongate (in the male), and more than four times as long as the outer pair.

First primary flight-feather *shortest*, the tenth *longest;* secondary quills enormously developed, much longer than the primaries, the eighth and ninth being nearly twice as long as the first.

Sides of the face, throat, and fore-part of neck naked.

Tarsus *much longer* than the middle toe and claw.

I. THE ARGUS PHEASANT.   ARGUSIANUS ARGUS.

Phasianus argus, Linn. S. N. i. p. 272 (1766).

Argus giganteus, Temm. Pig. et Gall. ii. p. 410 (1813), iii. p. 678 (1815) ; Jardine and Selby, Ill. Orn. n. s. pls. 14 and 15 (1837); Elliot, Monogr. Phasian. i. pl. 11 (1872); Hume and Marshall, Game Birds Ind. i. p. 99, pl. (1878) ; Sclater, P. Z. S. 1879, p. 115, pls. vii. and viii. fig. 1.

Argus pavoninus, J. E. Gray, Ill. Ind. Zool. i. pl. 36 (1830-32).

Argusianus argus, Gould, B. Asia, vii. pl. 52 (1883); Ogilvie-Grant, Cat. B. Brit. Mus. xxii. p. 363 (1893).

(*Plate XXIX.*)

Adult Male.—A short black crest ; general colour above black, chequered and mottled *with buff and rufous ;* lower back and rump *buff*, with rounded black spots ; long middle tail-feathers whitish-buff with kidney-shaped black spots and blotches ; primary quills beautifully patterned, and ornamented with close-set rows of black and rufous spots ; a rufous-brown band, finely dotted with white, on the basal part of the inner web *only*, divided from the dull-blue shaft by a yellow line regularly barred with black ; secondary quills equally wonderful in their markings, and having the outer webs decorated with a row of large ocelli, gradually increasing in size towards the extremity

ARGUS PHEASANT

of the feathers; each of these eyes yellowish in the centre, shading into white on the one side and reddish-brown on the other, and bounded by a black band; under-parts black with wavy bars and markings of chestnut and buff. Naked skin on sides of head and throat dark blue. Total length, 72 inches; wing to the end of the primary quills, 19; to the end of the secondaries, 34; tail, 50; tarsus, 4·4.

Adult Female.—Neck *chestnut*, slightly mottled with black, shading into reddish-buff on the mantle, which is thickly mottled with black; lower back bright buff, barred and mottled with black; wing-coverts and secondary quills black, thickly covered with buff hieroglyphics; primary quills chestnut, irregularly marked with black; under-parts *rufous*, finely mottled with black. Total length, 30 inches; wing to the end of the primary quills, 13; to the end of the secondaries, 15; tail, 12·5; tarsus, 4.

Range.—Laos Mountains, Siam, South Tenasserim, the Malay Peninsula, and Sumatra.

Habits.—The late Mr. W. R. Davison, who had exceptional opportunities of studying the habits of the Argus Pheasant, gives the following excellent account:—"They live quite solitarily, both males and females. Every male has his own drawing-room, of which he is excessively proud, and which he keeps scrupulously clean. They haunt exclusively the depths of the evergreen forests, and each male chooses some open level spot —sometimes down in a dark, gloomy ravine entirely surrounded and shut in by dense crane-brakes and rank vegetation; sometimes on the top of a hill where the jungle is comparatively open—from which he clears all the dead leaves and weeds for a space of six or eight yards square, until nothing but the bare clean earth remains, and thereafter he keeps this place scrupulously clean, removing carefully every dead leaf or twig that may happen to fall on it from the trees above.

"These cleared spaces are undoubtedly used as dancing-grounds, but personally I have never seen a bird dancing in them, but have always found the proprietor either seated quietly in, or moving backwards and forwards slowly about, them, calling at short intervals, except in the morning and evening, when they roam about to feed and drink. The males are always to be found at home, and they roost at night on some tree quite close by.

"They are the most difficult birds I know of to approach. A male is heard calling, and you gradually follow up the sound, taking care not to make the slightest noise, till at last the bird calls within a few yards of you, and is only hidden by the denseness of the intervening foliage. You creep forward, hardly daring to breathe, and suddenly emerge on the open space, but the space is empty; the bird has either caught sight of, or heard, or smelt you, and has run off quietly. They will never rise, even when pursued by a dog, if they can possibly avoid it, but run very swiftly away, always choosing the densest and most impenetrable part of the forest to retreat through. When once the cleared space is discovered it is merely the work of a little patience to secure the bird by trapping it. The easiest way is to run a low fence of cut scrub round the spot, leaving four openings just sufficiently wide to enable the bird to pass through, and in these openings to place nooses fastened to the end of a pliant sapling, which is bent and kept down by a catch. This is the usual way, and the one I adopted to secure most of my specimens, as I found it as difficult to shoot, as it was easy to trap, them. The natives, however, have other ways of securing them, all dependent on taking advantage of the bird's idiosyncrasy about keeping its home clean. . . .

"The males are not at all quarrelsome, and apparently never interfere with each other, though they will answer each other's calls. The call of the male sounds like '*how-how*,' repeated ten or a dozen times, and is uttered at short intervals when the bird is in its clearing, one commencing, and others in the neighbourhood answering. The report of a gun will set every

.male within hearing calling, and on the least alarm or excitement, such as a troop of monkeys passing overhead, they call.

"'The call of the female is quite distinct, sounding like '*howowoo, how-owoo,*' the last syllable much prolonged, repeated ten or a dozen times, but getting more and more rapid until it ends in a series of 'owoos' run together. Both the call of male and female can be heard to an immense distance, that of the former especially, which can be heard at the distance of a mile or more. Both sexes have also a note of alarm,—a short, sharp, hoarse bark.

"'The female, like the male, lives quite solitarily, but she has no cleared space, and wanders about the forest apparently without any fixed residence. The birds never live in pairs, the female only visiting the male in his parlour for a short time.

"'The food consists chiefly of fallen fruit, which they swallow whole, especially one about the size and colour of a prune, which is very abundant in the forests of the south; but they also eat ants, slugs, and insects of various kinds. These birds all come down to the water to drink, at about 10 or 11 a.m., after they have fed, and before they, or at any rate the males, return to their parlours."

Nest.—Said to be rudely constructed on the ground in some dense cane-brake. According to natives, the breeding-season continues all the year round, except during the depth of the rains.

Eggs.—Said to be seven or eight in number, white or creamy, minutely speckled with reddish-brown like a Turkey's. Measurements, 2·6 by 1·9 inches.

II. GRAY'S ARGUS PHEASANT. ARGUSIANUS GRAYI.

Argus grayi, Elliot, Ibis, 1865, p. 423; id. Monogr. Phasian. i. p. xviii. pl. 12 (1872).

Argusianus grayi, Ogilvie-Grant, Cat. B. Brit. Mus. xxii. p. 365
    (1893).

Adult Male.—Differs chiefly from the male of *A. argus* in
having the mantle and wing-coverts black, mottled with *white*
and rufous ; the lower back and rump *rufous-buff ;* the neck
and upper chest brighter rufous, with yellowish shaft-stripes ;
and the sides and flanks more or less mottled with white.   Size
smaller.   Total length, 60 inches ; wing to the end of the
primary quills, 17; to the end of the secondaries, 33 ; tail, 42 ;
tarsus, 4.

Adult Female.—Differs from the female of *A. argus* in having
the neck *rust-red ;* and the under-parts *sandy-brown*, but
slightly mixed with rufous and finely mottled with black.
Total length, 29 inches; wing to the end of the primary
quills, 12·6; to the end of the secondaries, 14 ; tail, 11·8 ;
tarsus, 3·6.

Range.—-Borneo.

Eggs.—Pale creamy-white, rather finely pitted all over with
reddish-brown.   Measurements, 2·6 by 1·85 inches.

III. WOOD'S ARGUS PHEASANT.   ARGUSIANUS BIPUNCTATUS.

Argus bipunctatus, Wood, Ann. Mag. N. H. (4) viii. p. 67
    (1871) ; Elliot, Monogr. Phasian. i. pl. 13 [part of pri-
    mary quill] (1872).
Argusianus bipunctatus, Ogilvie-Grant, Cat. B. Brit. Mus. xxii.
    p. 366 (1893).

This species is still known only from a portion of a primary
wing-feather from a male bird, which is now preserved in the
British Museum, to which it was presented by Mr. Edward
Bartlett.   It is so perfectly distinct both in its markings and
in the shape of the shaft of the quill, from either of the above
species, that one can have no hesitation in recognising it as
belonging to a distinct species, in spite of the somewhat frag-
mentary evidence.

The general colour of this feather is similar to that of a primary quill of *A. argus* and *A. grayi*, but there is a reddish-brown band minutely dotted with white *on the outer as well as the inner web*. These bands extend over a large proportion of both the webs; the shaft is remarkably narrow and attenuated.

Range.—Unknown.

THE CRESTED ARGUS PHEASANTS. GENUS RHEINARDTIUS.

Rheinardtius, Oustal. Bull. Soc. Philom. (7) vi. p. 254 (1882).

Type, *R. ocellatus* (Bp.).

Tail composed of *twelve feathers*, the middle pair being *enormously elongate* (in the male), the second pair considerably shorter, and the outermost pair about one-fifth of the length of the middle pair.

First primary flight-feather much shorter than the second; fourth about equal to the tenth; *fifth and sixth sub-equal and longest*.

Secondary flight-feathers *not* longer than the primaries.

Sides of the head naked; crown feathered, an erect crest of hairy feathers covering the nape.

Tarsus rather *shorter* than the middle toe and claw.

1. RHEINARDT'S ARGUS PHEASANT. RHEINARDTIUS OCELLATUS.

Argus ocellatus, Verr.; Bonap. C. R. xlii. p. 878 (1856)*; Elliot, Monogr. Phasian. i. pl. 13 (1872) [tail feathers only].

Rheinardtius ocellatus, Oustal. Ann. Sci. Nat. (6) xiii. Art. 12 (1882); id. N. Arch. Mus. (2) viii. p. 256, pl. ii. (1885).

* Described from a few tail-feathers in the Paris Museum.

Argus rheinardti, Maingonnat, Bull. Soc. Zool. France, vii. p. xxv. (1882).

Rheinardtius ocellatus, Ogilvie-Grant, Cat. B. Brit. Mus. xxii. p. 367 (1893).

Adult Male.—General colour dark brown, mixed here and there, especially on the under-parts, with rufous, and thickly covered with small white spots and markings; upper tail-coverts and the enormously elongate middle pair of tail-feathers grey, thickly covered with large spots and markings of chestnut, the spots on each side of the shaft with black central rings and smaller rounded dots of white; outer tail-feathers reddish-brown, thickly covered with round white spots surrounded by rings of black. Total length, about seven feet, wing, 13'5 inches; tail, 5 feet; tarsus, 3'5 inches; middle toe and claw, 3'7.

Adult Female.—Crest smaller than in the male; general colour above umber-brown, transversely mottled with black and buff, these markings being stronger on the secondaries and tail-feathers; below brown, finely mottled with black. Total length, about 31 inches; wing, 11'5; tail, 14'5; tarsus, 3'4.

Range.—Mountains in the interior of Tonkin.

This pheasant, still one of the rarest in collections, was first described in 1856 from some tail-feathers in the Paris Museum. Nothing more was known of it till 1882, when several pairs were obtained by the French during the Tonkin War, and in course of time found their way to Paris. Of these, the British Museum was fortunate enough to secure a fine adult pair, which were subsequently beautifully mounted by Mr. Pickhardt, the well-known taxidermist, and may now be seen exhibited in a case along with the Common Argus-Pheasant at the Natural History Museum. Nothing is recorded about this bird's habits, but they probably do not differ greatly from those of the Common Argus.

THE PEA-FOWL. GENUS PAVO.

Pavo, Linn. S. N. i. p. 267 (1766).

Type, *P. cristatus*, Linn.

Tail long and wedge-shaped, composed of twenty feathers; upper tail-coverts enormously developed in the male, forming the "train."

First primary flight-feather *much shorter* than the tenth; fifth somewhat the longest.

An elevated crest of feathers.

Tarsus armed in the male with a short stout spur.

Only two species are known.

1. THE COMMON PEA-FOWL. PAVO CRISTATUS.

Pavo cristatus, Linn. S. N. i. p. 267 (1766); Elliot, Monogr. Phasian. i. pl. 3 (1872); Hume and Marshall, Game Birds Ind. i. p. 81, pl. (1878); Oates, ed. Hume's Nest and Eggs, Ind. B. iii. p. 405 (1890); Ogilvie-Grant, Cat. B. Brit. Mus. xxii. p. 368 (1893).

Adult Male.—Crest of erect *naked* shafts, *with* fan-shaped bluish-green *plumes at the extremity;* the lesser and median wing-coverts, shoulder-feathers and inner secondary quills *pale buff, barred* and mottled *with black*, slightly glossed with green; primary quills and their coverts *pale chestnut;* thighs *whitish buff.* Naked skin on sides of head livid white. Total length to the end of tail, 35 inches; to the end of upper tail-coverts or train, 78; wing, 17·5; tail, 19·5; tarsus, 5·5.

Adult Female.—Terminal plumes of crest-feathers mostly chestnut edged with golden-green; feathered parts of head mostly dark chestnut, paler on the neck; mantle golden-green; rest of upper-parts brown, indistinctly mottled with buff; wing-coverts more coarsely mottled with buff and black; throat and part of neck white; chest brownish-black, edged with green and buff; under-parts buff, browner on the belly. Total length, 32 inches; wing, 16; tail, 13; tarsus, 4·8.

Habits.—Mr. A. O. Hume writes :—"An Indian bird *par excellence*, the Common Pea-Fowl, though widely spread throughout India proper, does not normally extend elsewhere, except into Ceylon and Assam.

"Even within these limits it is not universally distributed, as it affects water and cultivation, and in no way shuns the abodes of men. But there may be too much water, cultivation, and population to suit its taste.

"As a rule, the Pea-Fowl is not a bird of high elevations. On the Nílgiris I know it occurs as high as 5,000 feet at Cook's Hill, on the north-east slopes of those mountains, and it may even, as Jerdon says so, though I have been unable to verify this, occur up to 6,000 feet, but it does not, I believe, ascend the Pulneys, or the Ceylon Hills, to elevations of above 3,000 feet ; and in the Himalayas, though in the river valleys it penetrates, as in Central Gahrwal, far into the hills, it is rarely seen above 2,000 feet. Broken and jungly ground, where good cover exists, near water on the one hand, and cultivation on the other, is the favourite resort of the Pea-Fowl, and, wherever this favourable combination exists within the limits indicated, there the Pea-Fowl is sure to abound. Canals, with their grass- and tree-clad banks, are, in Upper India, pet abiding places of the species. . . .

"But it is not only in such seemingly suitable localities that this species thrives amazingly ; it is to be seen almost throughout Rajputana. In and about the rocky and semi-desert tracts, for instance, in which lie Jeypore, and the more ancient capital of that state, Umber, myriads of Pea-Fowl are to be met with. Everywhere throughout Upper India a certain superstitious reverence attaches to the Pea-Fowl, and the mass of the population more or less dislike their slaughter ; but, in these native states, the prohibition is absolute, and no man, native or European, can or does molest them, though tigers and leopards, if the people speak truly, are less amenable to authority. . . .

"The Pea-Fowl is at times omnivorous, and land-shells, insects of all kinds, worms, small lizards, and even tiny frogs may be found in their crops, but by choice I think they feed on grain and tender juicy shoots of grass and flower-buds, and I have scores of times examined their stomachs without finding a trace of anything else, although, had they been so minded, animal food of all kinds abounded round them.

"Where numerous, they do much damage to cultivation, and, being excessively fond of the buds of trees, are also very destructive to young plantations."

In Colonel Tickell's delightful account of this species we read :—"Pea-Fowl roost at night on high trees. The highest they can get in the jungle they inhabit; but they select the lowest branches for their perch. They are rather late in roosting, and I have heard them flying up to their berths long after sunset, and when the Night-Jars had been for some time abroad, flitting over the dusky jungle. The cock-bird invariably leads the way, rising suddenly from the brushwood near the roosting-tree, with a loud 'kok-kok-kok-kok,' and being presently followed by his harem—four or five hens. If marked to their roosting-place, and if it be a clear moonlight night, they may be easily shot, for, not knowing where to go, they will frequently remain on the tree till fired at two or three times. When forced to quit, they fly towards the ground, and pass the rest of the night as well as they can, sometimes falling a prey to leopards or wild cats. If there are hills in the jungle, the Pea-Fowl select some prominent tree on the top, or half-way up. In the Nílgiris and other mountain regions in Southern India, says Jerdon, this bird ascends to the height of 6,000 feet above the sea ; but in Sikhim (Darjiling) and other parts of the Himalaya, not higher than 2,000 feet. . . ." Colonel Tickell continues :—"In the months of December and January, the temperature in the forests of Central India, especially in the valleys, is very low, and the cold, from sudden evaporation, intense at sunrise. The Pea-Fowl in the forests may be observed at such times still roosting, long after the sun has risen above the horizon. As the mist rises off the valleys, and, gathering into little clouds, goes rolling up the hill-sides till lost in the ethereal blue, the Pea-Fowl descend from their perch on some huge símal or sál tree, and, threading their way in silence through the underwood, emerge into the fields, and make sad havoc with the channa, urad (both vetches), wheat, or rice. When sated, they retire into the neighbouring thin jungle, and there preen themselves, and dry their bedewed plumage in the sun. The cock stands on a mound, or a fallen trunk, and sends forth his well-known cry, 'pehaun-pehaun,' which is soon answered from other parts of the forest. The hens ramble about, or lie down dusting their plumage, and so they pass the early hours while the air is still cool, and hundreds of little birds are flitting and chirruping about the scarlet blossoms of the 'palás,' or the 'símal.' As the sun rises, and the dewy sparkle on the foliage dries up, the air becomes hot and still, the feathered songsters vanish into shady nooks, and our friends the Pea-Fowl depart silently into the coolest depths of the forest, to some little sandy stream canopied by verdant boughs, or to thick beds of reeds and grass, or dense thorny brakes overshadowed by mossy rocks, where, though the sun blaze over the open country, the green shades are cool, and the silence of repose unbroken, though the shrill cry of the Cicada may be heard ringing faintly through the wood.

"These birds cease to congregate soon after the crops are off the ground. The pairing-season is in the early part of the hot weather. The Peacock has then assumed his full train, that is, the longest or last rows of his upper tail-coverts, which he displays of a morning, strutting about before his wives. These strange gestures, which the natives gravely denominate the Peacock's nautch, or dance, are very similar to those of a Turkey-cock, and accompanied by an occasional odd shiver of the quills, produced apparently by a convulsive jerk of the abdomen. The same thing occurs in a Turkey-cock, a little start and a puff, and a short run forward, as if something had exploded unpleasantly close behind him. These are all blandishments, we are told, to allure the female, and doubtless have a most fascinating effect."

Mr. Sanderson writes as follows :—" Pea-Fowl run very fast, but the old cocks, burthened with tails six feet in length, are poor flyers, and I have frequently seen my men run them down during the hot hours of the day by forcing them to take two or three long flights in succession, in places where they could be driven from one detached patch of jungle to another.

" The old cocks are in full plumage from June to December, and then cast their trains.

"Pea-Fowl are, perhaps, the most wary of all jungle creatures. In beating for large game, where the sportsmen are posted ahead in trees, their presence may pass undetected by other animals, but rarely by Pea-Fowl."

A well-known variety of the Common Pea-Fowl is that described by Latham [Gen. Hist. viii. p. 114 (1823)] under the name of the Black-Shouldered Peacock, and subsequently named *Pavo nigripennis* by Dr. Sclater (P. Z. S. 1860, p. 221). This form differs from the male of typical *P. cristatus* in having the lesser and median wing-coverts, shoulder-feathers, and inner secondary quills *brownish-black*, more or less glossed with purple and edged with green, with only traces of buff mottlings on some of the secondaries, the primary quills and their coverts being *black along the shaft and margin of the inner web*, and the thighs black.

Although this variety closely resembles the male hybrids between *P. cristatus* and the following species *P. muticus*, it has been clearly shown that it arises independently in flocks of Common Pea-Fowl which have been pure-bred for many years, and there can be no doubt that it is merely a sport of nature, possibly due to atavism or reversion to the ancestral type, from which both the Common and Burmese Pea-Fowl have sprung. The male is well figured by Mr. Elliot [Monogr.

Phasian. i. pl. 4 (1872)], but the figure supposed to represent
the female of the Black-Shouldered form is merely a pale
variety. Albinos and pale cream-coloured forms are occasionally met with in a perfectly wild state, but the so-called *Pavo
nigripennis* has at present only been observed among birds in
captivity. Sportsmen should look carefully at any male Pea-
Fowl they may chance to shoot from time to time, as it would
be extremely interesting to know if this form ever occurs among
wild Indian birds.

Nest.—A hollow scratched by the hen in the ground and
lined with a few leaves and twigs among thick grass or dense
bushes.

Eggs.—Ten to fifteen in number ; broad ovals, varying in
colour from whitish to pale buff ; shell smooth and strong,
pitted all over. Average measurements, 2·74 by 2·05 inches.

II. THE BURMESE PEA-FOWL. PAVO MUTICUS.

Pavo muticus, Linn. S. N. i. p. 268 (1766) ; Elliot, Monogr.
   Phasian. i. pl. 5 (1872); Hume and Marshall, Game Birds
   Ind. i. p. 94, pl. (1878) ; Ogilvie-Grant, Cat. B. Brit. Mus.
   xxii. p. 371 (1893).
Pavo spiciferus, Shaw and Nodder, Nat. Misc. xvi. pl. 641 ;
   Vieill. Gal. Ois. ii. p. 14, pl. 202 (1825).
Pavo javanicus, Horsf. Tr. Linn. Soc. xiii. p. 185 (1822).
Pavo aldrovandi, Wilson, Ill. Zool. pls. xiv. and xv. (1831).
Pavo spicifer, Schinz, Nat. Vög. p. 150, pl. 73 (1853).

Adult Male.—Easily distinguished from the male of *P. cristatus*
by having the rather long erect crest of normally developed
feathers *webbed to the base of the shaft ;* the feathers of the
back copper-coloured, surrounded by golden-green and margined with black ; the wing-coverts and shoulder-feathers *black*,
glossed with purplish-blue and edged with green ; the thighs
black, glossed with green, and the naked skin round the eyes
bluish-green, on the cheeks *chrome-yellow*. Total length to end

of tail, 51 inches; to end of upper tail-coverts, 32·5; wing, 19·1; tail, 22·5; tarsus, 6·6.

Adult Female.—Differs from the male in having the back, shoulder-feathers, and rump brownish-black with but little green gloss, and indistinctly mottled with buff; the upper tail-coverts extending nearly to the end of the tail, and of a golden-green hue irregularly barred with buff. Total length, 44 inches; wing, 16·7; tail, 16; tarsus, 5·4.

Range.—Indo-Chinese countries, extending north to Chittagong, east through Siam to Cochin China, and southwards through the Malay Peninsula. It is also found in Java, and is stated by some of the earlier writers to inhabit Sumatra, but its occurrence there is doubtful.

Habits.—In many respects, as regards habits, food, and modes of life, the Eastern bird closely resembles the Indian one, but there is this essential difference, at any rate everywhere within our limits, that the Eastern bird is never found in thousands, throughout unbroken stretches of country, a hundred or more miles in length, as the Indian bird is, but only in small colonies in isolated spots, where one may often travel fifty or a hundred miles before coming across another colony.

Like its congeners, it moves about feeding, morning and evening, advancing into fields, if there happen to be cultivation near at hand at these times, and retreating during the day to dense cover. At night, of course, it roosts upon trees, and its call-note, like that of the Indian bird, which it closely resembles, is a harsh "mew, mew, mew," which one might fancy to be the cry of some gigantic tom-cat in distress. (*Hume.*)

To quote Colonel Tickell:—"The habits of *Pavo muticus* are so similar to those of its congener as scarcely to admit of separate description; but I should say it was a still more strictly sylvan or forest-haunting bird. Cultivation does not appear to entice it far from its leafy fastnesses, as it does the Bengal species, and it is, in consequence, more secluded, wilder, and more

G 2

difficult of approach, besides being far less numerous. I have
never seen more than three or four of the Burman Pea-Fowl
together, whereas the Bengal birds unite in flocks of thirty,
forty, or fifty. It haunts the thickest jungle, whether on level
ground or on the sides of small hills, and is frequently found
in the masses of elephant-grass which so commonly skirt the
smaller brackish creeks and nullas of Arakan. A specimen
with a full train is seldom seen except in the beginning of
the rains, which is the season of courtship. About August
they moult, drop their long ocellated tail-coverts, and assume
the simpler green-barred ones. The train appears again
in the succeeding March or April; but the moulting of this
bird appears to be irregular, and I have seen cock-birds with
fine flowing trains in January and February. The hen incubates in the rains, but at uncertain periods; the young, just
hatched, have been brought to me at Moulmein at different
times, from August till January."

Eggs.—Cannot be distinguished from those of *P. cristatus.*

THE BLACK GUINEA-FOWLS. GENUS PHASIDUS.

Phasidus, Cassin, Proc. Acad. Philad. 1856, p. 322.

Type, *P. niger,* Cassin.

Head and neck naked, with the exception of a band of
feathers along *the middle of the head,* commencing at the
base of the bill, and a few small scattered plumes on the
neck.

Tail moderately long and rounded, probably composed of
fourteen feathers.* Upper tail-coverts *about two-thirds* of the
length of the tail.

First primary flight-feather considerably *shorter than* the
second, which is about equal to the tenth; fourth slightly the
longest.

* Both the examples of this rare bird in the British Museum have imperfect tails.

Tarsus in male armed with a short blunt spur.
Only one species is known.

I. THE BLACK GUINEA-FOWL. PHASIDUS NIGER.

Phasidus niger, Cassin, Proc. Acad. Philad. 1856, p. 322 ; id. J.
Ac. Philad. (2) p. 7, pl. 3 (1858); Elliot, Monogr. Phasian.
ii. pl. 36 (1872) ; Rochebr. Act. Soc. Linn. Bord. xxxviii.
p. 356, pl. xxii. (1884); Ogilvie-Grant, Cat. B. Brit. Mus.
xxii. p. 373 (1893).

Adult Male.—A band of feathers along the middle of the
head, black ; general colour and rest of plumage blackish-
brown, finely mottled with dark brown. Naked skin of head
and neck Naples-yellow, shading into orange-yellow on the
throat and lower parts of the neck. Total length, 16·5 inches ;
wing, 8·8 ; tail,* 5·5 ; tarsus, 2·7.

Adult Female.—Similar to the male, but without spurs.

Range.—West Africa, from Cape Lopez to Loango.

Habits.—The Black Guinea-Fowl is one of the rarest of the
Game-Birds. Even in the British Museum collection there
are only two examples of it, and neither of these are per
fect specimens, the middle tail-feathers in both being absent.
Mr. Du Chaillu, the original discoverer of this remarkable bird,
gives the following brief account :—" One day I went out hunt-
ing by myself, and, to my great joy, shot another new bird, a
black wild fowl, one of the most singular birds I have seen in
Africa. . . . The head, where it is bare, is in the female
of a pink hue, and in the male of a bright scarlet.† . . . .
When I saw this bird for the first time in the woods I thought
I saw before me a domestic chicken. The natives have noticed
the resemblance, too, as their name for it shows, *couba iga*,

---

* Cassin gives the length of the tail as 6·0 inches.

† It will be noted that the colours of the naked skin here given do not
agree with those given in the description above. I am unable to say which
is correct.

signifying 'wild fowl.' Wild they are, and most difficult to
approach, and also rare, even in the forests where they are at
home. They are not found at all on the sea-coast, and do not
appear until the traveller reaches the range of fifty or sixty
miles from the coast. Even there they are so rare, that though
I looked out for them constantly I killed but three in all my
expeditions. They are not gregarious, like the Guinea-Fowl,
but wander through the woods, a male and one or, at most,
two females in company. They are very watchful, and fly off
to retreats in the woods at the slightest alarm."

THE TURKEY-LIKE GUINEA-FOWLS. GENUS AGELASTES.

Agelastes (Temm.), Bonap. P. Z. S. 1849, p. 145.

Type, *A. meleagrides*, Bonap.

Skin of head and greater part of neck *naked*, or with only a
few minute scattered plumes.

Tail rather long and rounded, composed of *fourteen* feathers,
the outer pair being not much shorter than the middle pair.
Upper tail-coverts *about two-thirds* of the length of the middle
pair of tail-feathers.

First primary flight-feather *much shorter* than the second,
which equals the tenth; fifth to seventh sub-equal and longest.

Tarsus in the male armed with a short stout spur.

Only one species is known.

I. THE TURKEY-LIKE GUINEA-FOWL. AGELASTES MELEAGRIDES.

Agelastes meleagrides (Temm.), Bonap, P. Z. S. 1849, p. 145 ;
     Schleg. Handl. Dierk. 1857, Vog. fig. 57 ; id. Dierentuin,
     1872, p. 220, cum fig. ; Elliot, Monogr. Phasian. ii. pl. 37
     (1872) ; Ogilvie-Grant, Cat. B. Brit. Mus. xxii. p. 374
     (1893).

Adult Male.—Lower-neck, chest, and upper mantle, *white ;*
rest of plumage black, finely mottled with white, primary flight-
feathers edged with whitish-grey ; naked skin of head *red,* darker

on the hind-neck, that of the lower-neck milky-white. Total length, 19 inches; wing, 8·7; tail, 6; tarsus, 2·9.

Adult Female.—Similar to the male, but devoid of spurs.

Range.—West Africa, from Liberia to Gaboon.

Like the Black Guinea-Fowl, this is one of the rarest birds in European collections, and up to the present time nothing is known regarding its habits. Mr. Büttikofer obtained several examples during his expeditions to Liberia, but these were bought from natives, who trap them in the bush-paths. According to native report, the Turkey-like Guinea-Fowl, though well-known, is very rare everywhere.

THE HELMETED GUINEA-FOWLS. GENUS NUMIDA.

Numida, Linn. S. N. i. p. 273 (1766).

Type. *N. meleagris*, Linn.

Head and neck *naked ;* a more or less elevated *bony helmet* covering the top of the head; a pair of wattles situated, one on each side, behind the angles of the gape.

Tail fairly long and somewhat rounded, composed of *sixteen* feathers, the middle pair being rather longer than the outer pair. Upper tail-coverts *almost extending to the end* of the tail.

First primary flight-feather *shorter than* the second, which is about equal to the tenth ; fifth slightly the longest.

In all the species the general colour of plumage is black, *spotted with white*[*]; the outer secondary quills *not white* on the outer web.

Sexes similar in plumage.

I. THE COMMON HELMETED GUINEA-FOWL. NUMIDA MELEAGRIS.

Numida meleagris, Linn. S. N. i. p. 273 (1766); Gray, Gen. Birds, iii. p. 501, pl. 128, fig. 2 (1845); Elliot, Monogr. Phasian. ii. pl. 39 (1872); Ogilvie-Grant, Cat. B. Brit. Mus. xxii. p. 375 (1893).

---

[*] The white spots appear to be absent on the upper-parts of *N. zechi*.

Numida rendallii, Ogilby, P. Z. S. 1835, p. 106; Fraser, Zool.
    Typ. pl. 62 (1841-2).
Numida maculipennis, Swains. B. W. Afr. ii. p. 226 (1837).
Numida marchei, Oustal. Ann. Sci. Nat. (6) xiii. Art. i. *bis*
    (1882); id. N. Arch. Mus. (2) viii. p. 305, pl. xiv. (1885).

Adult.—Easily distinguished from all the following species,
except *N. zechi*, by the *wide collar of vinous-grey* covering the
upper-part of the mantle and chest.   Naked skin on sides of
face and neck, chin and wattles red; rest of neck bluish.   Total
length, 25 inches; wing, 10·5; tail, 6; tarsus, 2·8; middle toe
(with claw), 2·2.

Range.—West Africa, extending from Senegambia southwards
through Ashantee to Gaboon; also met with in the Cape Verde
Islands, Annobon, and St. Thomas.

Habits.—Writing of this bird, which was introduced into the
Island of Jamaica more than two hundred years ago, Mr.
Gosse says:—" In a country whose genial climate so closely
resembled its own, and which abounded with dense and
tangled thickets, the well-known wandering propensities of
the Guinea-Fowl would, no doubt, cause it to become wild
very soon after its introduction.   It was abundant in Jamaica,
as a wild bird, 150 years ago, for Falconer mentions it among
the wild game, in his amusing 'Adventures.'   I shall confine
myself to a few notes of its present habits, which are, in all
probability, those of its original condition.

" The Guinea-Fowl make themselves only too familiar to the
settlers by their depredations in the provision-grounds.   In the
cooler months of the year they come in numerous coveys from
the woods, and, scattering themselves in the grounds at early
dawn, scratch up the yams and cocoes.   A large hole is dug
by their vigorous feet in very short time, and the tubers exposed, which are then pecked away, so as to be almost
destroyed, and quite spoiled.   A little later, when the planting
season begins, they do still greater damage, by digging up and

devouring the seed-yams and cococ-heads, thus frustrating the
hopes of the husbandman in the bud. 'The corn is no sooner
put in the ground than it is scratched out ; and the peas are
not only dug up by them, but shelled in the pod ' (*Dr. Cham*).
The sweet potato, however, *as I have been informed*, escapes
their ravages, being invariably rejected by them. To protect
the growing provisions, some of the negro peasants have re-
course to scarecrows, and others endeavour to capture the
birds by a common ' rat-gin ' set in their way. It must, however, be quite concealed, or it may as well be at home ; it is,
therefore, sunk in the ground, and lightly covered with earth
and leaves. A springe is useless, unless the cord be blackened
and discoloured so as to resemble the dry, trailing stem of
some creeper, for they are birds of extreme caution and sus-
picion. It is hence extremely difficult to shoot them, their
fears being readily alarmed, and their fleetness soon carrying
them beyond the reach of pursuit. But the aid of a dog, even
a common cur, greatly diminishes the difficulty. Pursuit by
an animal whose speed exceeds their own, seems to paralyse
them ; they instantly betake themselves to a tree, whence they
may be shot down with facility, as their whole senses appear
to be concentrated upon one subject, the barking cur beneath,
regarding whom with attent eyes and outstretched neck, they
dare not quit their position of defence. Flight cannot be
protracted by them, nor is it trusted to as a means of escape,
save to the extent of gaining the elevation of a tree ; the body
is too heavy, the wings too short and hollow, and the sternal
apparatus too weak, for flight to be any other than a painful
and laborious performance.

"Though savoury, and in high request for the table, the
Guinea-Fowl sometimes acquires an insufferably rank odour,
from feeding on the fetid *Petiveria alliacea ;* and is then
uneatable."

A supposed new species from Zanzibar described by Cabanis
under the name of *Numida orientalis* (cf. J. f. O. 1876, p. 210),

appears to be founded on a domestic variety of *N. meleagris* with abnormally developed wattles (one inch wide and half an inch long), at the angle of the gape.

N. meleagris has been known to cross with the Common Pea-Fowl (see Hocker, J. f. O. 1870, p. 152).

Nest.—Made in the midst of a dense tussock of grass.

Eggs.—About twelve are generally laid, and sometimes many more. Pale brownish or yellowish-buff, the whole shell thickly pitted with reddish-brown. The average measurement is 1·95 by 1·55 inch.

II. ZECH'S HELMETED GUINEA-FOWL. NUMIDA ZECHI.

Numida zechi, Reichenow, Orn. Monatsb. iv. p. 76 (1896).

Most nearly allied to *N. meleagris*, from which it appears to differ chiefly in having the feathers of the upper-parts pale brown or grey-brown, spotted with darker, black down the middle, and with very fine greyish-white streaks; the white ocelli being apparently wanting. Total length, 18·5 inches; wing, 11; tail, 6·4; tarsus, 3; middle toe and claws, 2·8.

Range.—West Africa ; Togo-land.

This form is evidently very closely allied to *N. meleagris*, and, if really distinct, apparently inhabits the same country. I have, as yet, had no opportunity of examining the type.

III. THE LARGE-HELMETED GUINEA-FOWL. NUMIDA CORONATA.

Numida coronata, Gray, List of Birds, pt. iii. Gall. p. 29 (1844); Elliot, Monogr. Phasian. ii. pl. 40 (1872) ; Ogilvie-Grant, Cat. B. Brit. Mus. xxii. p. 376 (1893) [part.; Eastern South Africa].

Adult.—Bony helmet long, high, and compressed (height, 1–1·5 inch, length, ·9*), sloping obliquely backwards; feathers

---

* The height is measured from the middle of the base to the apex ; the length, at the base of the helmet.

round the base of the neck more or less barred with narrow black and white bands; basal part of helmet red. Naked skin on the sides of head and neck pale blue. *Wattle pale blue, tipped with red.* Total length, about 20 inches; wing, 11·5; tail, 6·4; tarsus, 3·1; middle toe (with claw), 2·85.

Range.—Eastern South Africa, extending, so far as is at present known, from the shores of the Zambesi to Cape Colony.

Habits.—Until recently this fine Guinea-Fowl was considered identical with the East African Bird, which has now been separated under the name of *N. reichenowi.* It has also been confounded by many authors with the larger *N. marungensis* from Benguela and Marungu, and with *N. cornuta* from Damara-land. The habits of the Large-Helmeted Guinea-Fowl have never been very fully described, but appear to be very similar to those of the allied Western forms. Mr. Ayres tells us that in Natal they are gregarious, and generally found among scrubby bush on the borders of streams and rivers. They run very rapidly, and in open ground a person on foot would stand but a poor chance of running them down. In cover they lie very close indeed, and require a good dog to find them; when found, they will frequently fly up into the lower boughs of any convenient bush or tree. They are naturally very tame and easily domesticated, and may be seen thus at any farmstead; in some instances they come regularly to feed with the poultry.

Eggs.—Like those of *N. meleagris*, but considerably smaller and rounder in shape. Measurements, 1·7 by 1·4 inch.

IV. REICHENOW'S LARGE-HELMETED GUINEA-FOWL.
NUMIDA REICHENOWI.

Numida coronata, Ogilvie-Grant, Cat. B. Brit. Mus. xxii. p. 376 (1893) [part.; E. Africa].
Numida reichenowi, Ogilvie-Grant, Ibis, 1894, p. 535, fig. i.

Adult.—Distinguished from *N. coronata* by having the large

bony helmet set *vertically* on the head; the wattles at the gape
entirely crimson-red.   Total length, 20·5 inches; wings, 11·0;
tail, 6·0; tarsus, 3·1 ; middle toe (with claw), 2·65.

Range.—Eastern Africa, extending in the south to the
Pangani River, westwards to Kakoma, and northwards to
Teita, Ukambani, and the countries to the south of Victoria
Nyanza.

V. THE MARUNGU HELMETED GUINEA-FOWL.   NUMIDA MARUNGENSIS.

Numida coronata marungensis, Schalow, Zeitschr. ges. Orn. i.
p. 105 (1884).
Numida marungensis, Ogilvie-Grant, Cat. B. Brit. Mus. xxii. p.
377 (1893).

Adult.—Larger than any other species of *Numida ;* the bony
helmet *less elevated* and *stouter* than in either of the preceding
species (height, 0·7 inch; length, 1·2) ; the fine black and
white barring on the base of the neck *continued on to the chest.*
Naked skin on sides of head and throat flesh-colour.   Total
length, 27 inches; wing, 12·5 ; tail, 7·4; tarsus, 3·5; middle
toe (with claw), 3·15.

Range.—West Africa, extending from Benguela to Marungu,
west of Lake Tanganyika.

VI. THE DAMARA-LAND HELMETED GUINEA-FOWL.   NUMIDA CORNUTA.

Numida cornuta, Finsch and Hartl. Vög. Ost.-Afr. p. 569
(1870); Ogilvie-Grant, Cat. B. Brit. Mus. xxii. p. 378
(1893).
*Numida papillosa,** Reichenow, Orn. Monats. ii. p. 145 (1894);
Fleck, J. f. O. 1894, p. 320, fig.

---

* *Numida papillosa,* Reichenow [Orn. Monatsb. ii. p. 145 (1894)], from
Kalahari, is described as differing from *N. cornuta* in having a small warty
growth situated at the base of the culmen ; but this peculiarity is equally
to be found in birds from Damara-land and Mossamedes, which are un-

Adult.—Easily recognised by the shape of the bony helmet, which, though slightly curved backwards as in *N. coronata*, is *more elevated, about half* the breadth, and *nearly cylindrical* (height, 2·1 inches; length, 0·55); the white spotted plumage *continuous almost to the base of the bare neck*, only a few feathers at the base of the neck showing traces of bars. Naked skin of face clear blue, shading into purplish-blue on the neck; wattles blue, with scarlet tips. Total length, 21·5 inches; wing, 10·4; tail, 5·8; tarsus, 2·7; middle toe (with claw), 2·4.

Range.—Western South Africa; Great Namaqua-land, Damara-land, and Mossamedes.

Habits.—Mr. Andersson tells us that this Guinea-Fowl is the commonest Game-Bird in Damara and Great Namaqua Lands, being most abundant from the Orange River in the south to the Okavango in the north of those countries; and it is also very common in the lake regions. It is a highly gregarious bird, especially during the dry season, when it is not uncommonly found in flocks of several hundred individuals; and on one occasion he saw upwards of a thousand collected in one spot, which was one of the prettiest sights he had the good fortune to witness. These wonderful congregations usually occur in the immediate neighbourhood of waters of small extent; and it is quite evident that were such a mass of birds to make a simultaneous rush for the precious liquid, there would be much confusion, and comparatively few would be enabled to have their fill. But on the contrary they go to work most economically and judiciously, and it is very interesting to watch the process. The first comers enter the

doubtedly typical *N. cornuta*, while the size of the white spots on the chest and breast varies in individuals and is unimportant. Herr E. Fleck (J. f. O. 1884, p. 390) figures the head of *N. papillosa*, Reichenow, and compares it with *N. reichenowi*, Grant, but the head figured by him as *N. reichenowi* is that of *N. mitrata*, a very different species. This may be seen at a glance by the size and shape of the helmet and blue red-tipped wattles. In *N. reichenowi* the wattles are *entirely red*.

well or hole, as the case may be, and, rapidly and dexterously
taking their fill, they make their exit in a different direction, if
possible, from that by which they entered ; in the meanwhile,
the outsiders gradually and evenly approach, and the ring is
gradually narrowed by a steady progressive movement of the
whole. A batch of fresh-comers never attempt to force their
way amongst those which have previously arrived, but remain
quietly on the outside of the ring until their turn comes. This
Guinea-Fowl feeds on grass, seeds, and insects, but chiefly on
a small bulb, which is also eagerly sought for by all the gallinaceous birds. They rest during the heat of the day under
some mimosa, resuming their wanderings when the greatest
heat is passed. A flock of these birds is in general easily
discovered by their sharp, discordant, and metallic cries, something like a rapid succession of blows struck upon iron. They
have many enemies, and seek security at night by roosting in
tall mimosas.

Nest.—A slight rounded depression in the ground.

Eggs.—Fifteen to twenty in number; buffy-white or pale buff
colour, sometimes obscurely speckled with pale grey.

VII. PALLAS'S HELMETED GUINEA-FOWL.   NUMIDA MITRATA.

Numida mitrata, Pall. Spic. Zool. i. fasc. iv. p. 18, pl. 3 (1767);
    Elliot, Monogr. Phasian. ii. pl. 41 (1872); Meyer, Vog.-
    Skel. pt. x. pl. 99 (1886); Ogilvie-Grant, Cat. B. Brit.
    Mus. xxii. p. 378 (1893).
Querelea tiarata, Bonap. C. R. xlii. p. 876 (1856).
Numida tiarata, Hartl. Orn. Madagas. p. 68 (1861).
Numida reichenowi, Fleck (*nec* Ogilvie-Grant), J. f. O. 1894,
    p. 390, fig.

Adult.—Like *N. coronata*, but the bony helmet is *much
smaller* and *conical* in shape (height, 0·8 inch; length, 0·9).
Top of the head scarlet ; helmet paler ; naked skin on sides
of head and neck blue ; wattles blue, tipped with red. Total

length, 20 inches; wing, 10·5; tail, 6·2; tarsus, 3·2; middle
toe (with claw), 2·6.

Range.—East Africa, extending northwards from the Zambesi. It is also found in the Comoro Islands, Madagascar,
Rodriguez, and other islands into which it has probably been
introduced.

Habits.—This species is pretty generally distributed all over
Madagascar up to Beforona, but is most numerous along the
coast-line, where they may be found in the early morning
feeding among the ferns and brushwood on the outskirts of
the forests. (*Roch* and *E. Newton.*)

At Foule Point, according to Dr. Roch, this species is very
common, and he obtained eggs in November, several coveys
being observed between there and Nossi-bé. The natives
often hunt them with dogs, and he was told that the birds,
while endeavouring to conceal themselves from the latter, will
allow themselves to be taken in the hand, rather than fly or
run into the open. When "treed," they will remain with their
long necks stretched out in stupid astonishment, as long as the
d gs continue yelping underneath, paying no regard to their
dangerous pursuers, and thus affording an easy shot to the
native sportsman.

Eggs.—Much like those of the Common Guinea-Fowl, but
the shells are marked nearly all over with small blotches and
spots of pale light red or greyish-red. Average measurements
of two eggs only, 1·9 by 1·4 inch.

VIII. THE ABYSSINIAN HELMETED GUINEA-FOWL. NUMIDA PTILORHYNCHA.

Numida ptylorhyncha, Licht.; Less. Traité d'Orn. p. 498 (1831).
Numida ptilorhyncha, Rüpp. N. Wirbelth, p. 184 (1835–40);
    Gray, Gen. Birds, iii. p. 501, pl. 128 (1845); Elliot,
    Monogr. Phasian. ii. pl. 42 (1872); Ogilvie-Grant, Cat.
    B. Brit. Mus. xxii. p. 379 (1893).

Adult.—Easily distinguished from all the preceding species
by having *a bunch of horny bristles at the base of the upper mandible;* otherwise much like *N. mitrata;* inner webs of the
primaries uniform, or more or less spotted with white * ; upper
half of neck thinly covered with black feathers. Naked skin
on side of head, neck, and wattles blue; helmet and bristles
pale horn-colour. Total length, 19 inches ; wing, 11 ; tail, 6·4;
tarsus, 3·2 ; middle toe (with claw), 2·7.

Range.—Equatorial and North-east Africa. Tingasi, Shoa,
Abyssinia and Bogos-land to Suakim, Sennaar, and Kordofan.

Habits.—Mr. W. T. Blanford had many opportunities of studying the habits of this species during his travels in Abyssinia,
and met with it throughout the country from the sea-coast
itself to an altitude of at least 9,000 feet. He says that these
birds keep much to craggy places, especially to rocky valleys,
and often remain during the middle of the day on the sides
of the steep or precipitous hills. They feed either in open fields
or in woods amongst bushes, &c., in the morning and evening,
and roost at night on high trees, a grove of lofty junipers being
frequently selected for that purpose in the highlands.

Throughout the winter and spring the Guinea-Fowls remain
in large flocks, usually of 200 or 300 birds each. These sub-
divide into smaller flocks to seek food during the day, but keep
to one general tract of country, and unite again at night. Where
not pursued, they are not particularly wary, and but little difficulty is found in getting within gun-shot.

In July and August the flocks divide into pairs, two or three
of which are often found together, and the breeding-season
commences. At this time the birds never appear to collect in
large flocks ; he did not, however, happen to see any of the
roosting-places. He shot a female containing a fully-formed
egg on the 9th August.

---

* The absence or presence of white spots is apparently purely individual,
and has nothing to do with age or sex.

PLATE XXX.

Wyman & Sons Limited

BLACK - COLLARED   CRESTED   GUINEA - FOWL.

The young are probably hatched about the end of August or beginning of September, as they are full-grown by the end of the year.

The voice is very similar to that of the Domestic Guinea-Fowl. The food appears to consist to a larger extent of seeds and fruits than amongst the Partridges, insects being apparently but little sought after. In one instance three birds shot one morning near Halai had been feeding chiefly upon the small tubers or corms of the Quentee (*Cyperus esculentus*). Their crops also contained seeds and a few fragments of leaves, but amongst the three only one insect, a bug.

THE CRESTED GUINEA-FOWLS. GENUS GUTTERA.

Guttera, Wagler, Isis, 1832, p. 1225.

Type, *G. cristata* (Pallas).

A *well-developed crest of black feathers* covering the top of the head ; rest of the head and neck naked ; wattles at the angle of the gape small or well-developed ; a fold of skin at the back of the neck. Tail moderately long and somewhat rounded, composed of *sixteen* feathers. Upper tail-coverts *extending nearly to the end* of the tail-feathers.

First primary flight-feather *considerably shorter than* the second, which is about equal to the tenth ; fifth a trifle the longest.

Male *not* armed with spurs.

Plumage alike in both sexes. General colour black, spotted with *pale blue ;* outer webs of outer secondaries margined with *pure white.*

1. THE BLACK-COLLARED CRESTED GUINEA-FOWL. GUTTERA CRISTATA.

Numida cristata, Pallas, Spic. Zool. i. fasc. iv. p. 15, pl. 2 (1767); Elliot, Monogr. Phasian. ii. pl. 45 (1872).

Guttera cristata, Wagler, Isis, 1832, p. 1225; Ogilvie-Grant,
Cat. B. Brit. Mus. xxii. p. 381 (1893).

(*Plate XXX.*)

Adult.—Top of head covered by a full long crest of curling
black feathers; a *black collar confined to the base of the neck*,
hardly extending on to the chest; naked skin on rest of head
and neck cobalt-blue, except the chin and throat, which are
red. Total length, 20 inches; wing, 10·2; tail, 5·1; tarsus
3·0; middle toe (with claw), 2·4.

Range.—West Africa, extending from Sierra Leone to the
Gold Coast.

Habits.—Although this Crested Guinea-Fowl is a well-known
bird throughout Liberia, Mr. Büttikofer tells us that it is extremely difficult to obtain, on account of its extreme shyness
and its aptitude for hiding when met with. On open plains it
was never seen, being always killed in brushwood and high
forest when watching for Antelopes. Occasionally it was caught
in snares placed in narrow passages through the dense brushwood. Some of these Guinea-Fowls, kept in confinement by
Mr. H. T. Ussher during his residence in Fantee, appeared to
thrive well, and could probably be domesticated, but he tells us
that they proved a great nuisance amongst other birds, being of
a pugnacious disposition, especially when associated with the
domestic examples of the Common Helmeted Guinea-Fowl.

II. THE BLACK-CHESTED CRESTED GUINEA-FOWL.   GUTTERA
EDOUARDI.

Numida edouardi, Hartlaub, J. f. O. 1867, p. 36; id. Ibis,
1870, p. 444.
Numida verreauxi, Elliot, Ibis, 1870, p. 300; id. Monogr
Phasian. ii. pl. 44 (1872).
Numida sp., Sclater, P. Z. S. 1890, p. 86, pl. xii.
Guttera edouardi, Ogilvie-Grant, Cat. B. Brit. Mus. xxii. p. 382
(1893).

Adult.—Distinguished from the last species, *G. cristata*, in having the *black collar round the base of the neck extended over the whole chest*, which is usually more or less washed with chestnut. Naked skin on the sides of the head and neck dark purple, black round the eye, leaden-grey on the throat, and yellowish-grey round the back of the neck.* Total length, 20 inches ; wing, 10·4 ; tail, 5·1 ; tarsus, 3·1 ; middle toe (with claw), 2·7.

Range.—South Africa, extending from Natal and Zulu-land to the Zambesi at least as far west as the Victoria Falls, and perhaps westwards to Benguela [see Bocage, J. Ac. Lisb. No. xii. p. 275 (1871)], but this requires confirmation.

Habits.—Mr. Ayres tells us that he met with these fine Guinea-Fowls in the month of July at Durban, Natal, where they were being hawked about the town by Kafir hunters as birds for the table, the flesh being most uncommonly delicate and good. They frequent the dense bush immediately on the sea range, and are difficult to get. The best method is with dogs accustomed to hunt the bush, as the birds, when chased, take to the trees, and a good dog will bark till his master manages with much trouble to get to the spot through brambles, thorny bushes, and nettles innumerable, and then, if due care is taken to approach without noise, the birds may be potted from the tree, a flying shot being totally out of the question. This species appears to be very local, and we gather from the meagre notes at our disposal that its habits are extremely similar to those of the Helmeted Guinea-Fowls, to which it is closely related.

* The exact colours of the naked skin on the sides of the head and neck have not been satisfactorily ascertained in this species. The throat and fore-neck are said by Mr. Elliot to be bright red, the sides and back of neck light blue. It is quite possible that the colours become brighter in the breeding-season or vary with age, being brightest in the adults. The colours of these parts should be carefully noted as soon as the birds have been shot, for they rapidly change after death. It would be easy for sportsmen in South Africa to settle this point, which is one of considerable interest to ornithologists.

III. THE CURLY-CRESTED GUINEA-FOWL.   GUTTERA PUCHERANI.

Numida cristata, Shaw and Nodder (*nec* Pallas), Nat. Misc. pl. 757.

Numida pucherani, Hartl. J. f. O. 1860, p. 341 ; Elliot, Monogr. Phasian. ii. pl. 46 [naked skin incorrectly coloured] (1872).

Numida granti, Elliot, P. Z. S. 1871, p. 584 ; id. Monogr. Phasian, ii. pl. 43 [blue spots should be continuous to base of neck] (1872).

Numida ellioti, Bartlett, P. Z. S. 1877, p. 652, pl. lxv.

Guttera pucherani, Ogilvie-Grant, Cat. B. Brit. Mus. xxii. p. 383 (1893).

Adult.—Differs conspicuously from both the preceding species in having *no black collar, the blue spots being continuous right up to the base of the naked neck.* Crest full and *curly.* Naked skin on the head and throat red, on the back and sides of the head blue ; wattles red, very small ; fold of skin at the back of the neck well-developed. Total length, 20 inches ; wing, 10·8 ; tail, 5 ; tarsus, 3·6 ; middle toe (with claw), 2·7.

Range.—East Africa, extending from Zanzibar northwards to the Tana River and westwards into the interior.

We are told that this extremely handsome Guinea-Fowl is only met with in the forest along the banks of rivers, but practically nothing has been published respecting its habits.

IV. THE STRAIGHT-CRESTED GUINEA-FOWL.   GUTTERA PLUMIFERA.

Numida plumifera, Cassin, P. Ac. Philad. viii. p. 321 (1856) ; id. Journ. Ac. Philad. iv. p. 6, pl. 2 (1858); Elliot, Monogr. Phasian. ii. pl. 47 (1872).

Guttera plumifera, Ogilvie-Grant, Cat. B. Brit. Mus. xxii. p. 384 (1893).

Adult.—As in the last species, *G. pucherani*, the spotting is continuous to the base of the bare neck, but *the crest is composed of thin straight feathers growing upwards,* and the wattles are much larger (about half an inch in length). The naked skin on the head and neck probably bluish-purple, but no record of the colour of the soft parts taken from freshly-killed specimens is to be found. Total length, 20 inches ; wing, 9·6 ; tail, 4·5 ; tarsus, 3·0; middle toe (with claw), 2·5.

Range.—West Africa, extending from Cape Lopez to Loango. Recently recorded from the Cameroons.

Habits.—From the notes given by Mr. Du Chaillu, the original discoverer of this species, our sole information respecting the habits of this bird is derived. He writes:—"This bird is not found in the forests near the sea-shore, but is first met with, as I afterwards ascertained, about fifty miles east of Sangatanga. It is very shy, but marches in large flocks through the woods, where the traveller hears its loud voice. It utters a kind of 'quack,' hoarse and discordant, like the voices of other Guinea-Fowls. It avoids the path left by travellers; but its own tracks are met everywhere in the woods it frequents, as the flock scratch and tear up the ground wherever they stop. It is strong of wing, and sleeps by night on the tops of high trees, a flock generally roosting together on the same tree. When surprised by the hunter, they do not fly in a body, but scatter in every direction. Thus it is a difficult bird to get, and the natives do not often get a shot at it."

THE VULTURINE GUINEA-FOWLS. GENUS ACRYLLIUM.

Acryllium, Gray, List Gen. Birds, p. 61 (1840).

Type, *A. vulturinum* (Hardw.).

Head and upper part of neck naked, except a *horse-shoe shaped band* of feathers extending from the ear-coverts *round the nape.* Plumage of neck, chest, and mantle developed into *long, pointed hackles,*

Tail long, wedge-shaped, composed of *sixteen* feathers, the middle pair *much lengthened* and pointed.

First primary flight-feather shorter than the second, which is about equal to the ninth; sixth slightly longest.

Tarsus in male with four or five knobs.

Only one species is known.

I. THE VULTURINE GUINEA-FOWL.   ACRYLLIUM VULTURINUM

Numida vulturina, Hardw. P. Z. S. 1834, p. 52 ; Gould, Icon.
    Av. pl. 8 (1837); Elliot, Monogr. Phasian. ii. pl. 38
    (1872).
Acryllium vulturinum, Gray ; Ogilvie-Grant, Cat. B. Brit. Mus.
    xxii. p. 385 (1893).

Adult Male.—The band of velvety feathers on nape reddish-brown ; long hackles of neck, mantle, and chest with white middles, edged with black and margined with cobalt-blue ; breast and belly cobalt-blue, black down the middle ; sides and flanks washed with purple ; rest of plumage mostly black, minutely dotted and spotted with white.   Naked parts of head and neck cobalt-blue.   Total length, 30 inches ; wing, 12'2 ; tail, 11'3 ; tarsus, 4'1 ; middle toe (with claw), 3.

Adult Female.—Differs only in having no knob-like spurs on the tarsi, and in being rather smaller than the male.

Range.—East Africa, extending from the Pangani River northwards to Somali-land and westwards to Kilimanjaro.

Although this remarkably handsome species was first discovered in 1834, practically nothing is known regarding its habits.   Mr. H. C. V. Hunter, who met with it along the Useri River, tells us that it frequents dry red soil covered with thorny bushes; he also found it particularly plentiful on the Tana River, where, with the exception of the curly crested form (*Guttera pucherani*), it was the only Guinea-Fowl observed.

Eggs.—An egg laid in the Gardens of the Zoological Society of London is nearly white with a slight yellowish-buff tinge; the shell is rather smooth, with some gloss, and not very deeply pitted. Measurements, 1·95 by 1·55 inch.

THE TURKEYS. GENUS MELEAGRIS.

Meleagris, Linn. S. N. i. p. 268 (1766).

Type. *M. gallopavo* (Linn).

Head and neck *naked and wattled,* with only a few hair-like feathers; *an erectile fleshy process on the forehead.*

Tail broad and rounded, composed of *eighteen* feathers, the middle pair not much longer than the outer.

First primary flight-feather *about equal to* the tenth, fifth slightly the longest. Tarsus considerably longer than the middle toe and claw, and armed *in the males* with a large stout spur.

I. THE MEXICAN TURKEY. MELEAGRIS GALLOPAVO.

Meleagris gallopavo, Linn. S. N. i. p. 268 (1766); Ogilvie-Grant, Cat. B. Brit. Mus. xxii. p. 387 (1893).

Meleagris mexicana, Gould, P. Z. S. 1856, p. 61 ; Elliot, B. N. Amer. ii. pl. 38 (1869) ; id. Monogr. Phasian. i. pl. 32 (1872).

Meleagris gallopavo mexicana, Bendire, N. Amer. B. p. 116, pl. iii. fig. 15 [egg] 1892 [part].

Adult Male.—General colour of plumage black and dark copper-bronze, shot with fiery-green and purplish-bronze. A tassel-like bunch of long, coarse, black, hair-like feathers *on the middle of the breast ;* upper tail-coverts and tail-feathers broadly *tipped with white,* the latter *never* ornamented with metallic ocelli near the extremity, though the outer feathers have a slight metallic band across the middle of the sub-terminal black band; primary quills equally barred with dark

brown and white ; naked skin on head and neck pale crimson.
Total length, about 43 inches ; wing, 21 ; tail, 15·5 ; tarsus, 7.

Adult Female.—-Smaller and less brightly coloured than the
male, from which it also differs in having a narrow band of
feathers along the middle of the crown to the base of the
small erectile process on the forehead ; no bunch of hair-like
feathers on the breast; and the whole of the under-parts fringed
with white.   Total length, about 40 inches ; wing, 17·9 ; tail,
14·3 ; tarsus, 5·3.

Range.—Table-lands of North Mexico, Arizona, New Mexico,
and Western Texas.

Habits.—It is from the Mexican form that our domestic
breed of Turkeys has been derived.   There appears to be no
doubt that at the time of the Conquest these birds were
regularly reared in captivity by the Mexicans, and were
brought to Europe early in the sixteenth century either direct
from Mexico or from the West Indian Islands, where they had
been previously introduced.

The Mexican Turkey, according to Captain Bendire, is more
of a mountain-loving species than the Eastern bird, and is still
reasonably abundant in the wilder portions of Western Texas,
the Territories of New Mexico and Arizona, and is very
common in portions of Mexico.   He believes that this species
attains a greater size than *M. americana*, as he shot a specimen weighing twenty-eight pounds after being drawn, and
was informed that much heavier birds are occasionally
killed.   This he could readily believe, having seen tracks
of this species along the banks of the San Pedro River,
in Arizona, measuring between five and six inches in length,
and unquestionably made by a much larger bird than the one
he had killed.   Mr. Herbert Brown, of Tucson, Arizona,
remarks :—" Without knowing it positively, I am of the
belief that they raise two broods of young every season, as
I have seen almost all sizes in the mating-season (October),

when they congregate in large numbers in the cañons to feed on *ballotes*, a small bitter acorn, common to the cañons and parks of Southern Arizona and southward. I have seen their roosting-places at night, in sycamore (*Alisa*) trees ; I also saw one in an oak grove on the side of a hill, but they appear more to favour the cañons. On the head-waters of the Santa Dominga I have seen not less than fifty or sixty in a bunch, and Turkey, in those days, was a common camp fare. I have been told by Mexicans that coyotes catch Turkeys by running in circles under their roosting tree till the birds get dizzy with watching them, and fall down. I never saw it done, but have been assured that it is a fact."

"The mating-season," writes Captain Bendire, "commences, according to latitude, from March 1st to the middle of April, by which time some of the birds commence nesting.

"They are summer residents in the higher mountain ranges, reaching an altitude of from 8,000 to 10,000 feet, and retiring to the more sheltered cañons and the timbered river valleys in the late fall, congregating at such times in large flocks.

"The Mexican Turkey, like the Eastern species, is polygamous, and the female attends exclusively to the duties of incubation, which lasts about four weeks, the male not only not assisting, but, according to observations made by Lieut. J. M. F. Partello, Fifth Infantry, U.S. Army, often destroying the eggs and the tender young."

Nest.—No doubt similar to that of the sub-species *M. ellioti*, which is described below.

Eggs.—Creamy-white, spotted and dotted over the entire shell with reddish-brown. Average measurements, 2·7 by 1·9 inches.

SUB-SP. *a.* ELLIOT'S TURKEY. MELEAGRIS ELLIOTI.

Meleagris gallopavo ellioti, Sennett, Auk, 1892, p. 167, pl. iii,

Meleagris gallopavo mexicana, Bendire, N. Am. B. p. 116
(1892) [part].
Meleagris ellioti, Ogilvie-Grant, Cat. B. Brit. Mus. xxii. p. 388
(1893).

Adult Male.—Differs from typical *M. gallopavo* in having the
feathers of the rump, upper tail-coverts, and tail tipped with
pale rufous-buff, though it must be remarked that the colour
of these parts varies greatly in different individuals from the
same locality, one being nearly chestnut while another is pale
whitish-buff. The white bars on the primary quills are much
narrower.

Adult Female.—Like the female of *M. gallopavo*, but the tail-
coverts, tail, &c., tipped with rufous-buff as in the male.

Range.—Vera-Cruz and Tamaulipas, in Eastern Mexico, and
South-western Texas.

Nest.—Captain Gosse describes a nest of this sub-species
found in Southern Texas as being a coarse structure not very
deeply excavated, lined with grass, weeds, and leaves, and
placed in quite an open situation in open bushy country, but
well concealed by a few small bushes and bunches of growing
grass.

Eggs.—Eleven were found by Captain Gosse.

II. THE AMERICAN TURKEY. MELEAGRIS AMERICANA.

Meleagris americana, Bartram, Trav. p. 290 (1791); Ogilvie-
Grant, Cat. B. Brit. Mus. xxii. p. 389 (1893).
Meleagris palawa, Barton, Med. and Phys. J. ii. pt. i. pp.
163, 164 (1805).
Meleagris silvestris, Vieillot, N. Dict. d'Hist. Nat. ix. p. 447
(1817).
Meleagris fera, Vieillot, Gal. Ois. ii. p. 10, pl. 201 (1825).
Meleagris gallopavo, Bonap. (*nec* Linn.), Am. Orn. i. p. 79,
pl. ix. (1825); Aud. Orn. Biogr. i. pp. 1 and 33, pls. i. and

vi. (1831); id. B. Amer. v. p. 42, pls. 287, 288 (1842); Elliot, Monogr. Phasian. i. pl. 30 (1872); Bendire, N. Am. B. p. 112, pl. iii. fig. 14 (1892; part).

Adult Male.—Differs chiefly from *M. gallopavo* in having the lower rump, flanks, upper and under tail-coverts, and tail-feathers tipped with *deep chestnut-maroon*, and the white bars of the primary quills *rather* narrower, but as wide, or nearly as wide, as the dark interspaces.

Adult Female.—Similar to the male, but distinguished by having the feathers of the lower back and under-parts fringed with chestnut, those of the nape extending to the crown, and the pectoral tassel and spurs wanting.

Habits.—"The breeding range of the Wild Turkey, the largest and finest of our game-birds, is yearly becoming more and more restricted, and at the present rate of decrease its total extinction east of the Mississippi and north of the Ohio River is only a question of a few years. . . .

"The Wild Turkey is a resident wherever found. Numerous records attest the abundance of this magnificent game-bird throughout the Southern New England States in former years, and evidences of its existence have been found in Southern Maine.

"The Wild Turkey is essentially a woodland bird, and inhabits the damp and often swampy bottom lands along the borders of the larger streams, as well as the drier mountainous districts found within its range, spending the greater part of the day on the ground in search of food, and roosting by night in the tallest trees to be found. From constant persecution in the more settled portions of its range, it has become by far the most cunning, suspicious, and wary of all our game-birds, while in sections of the Indian Territory and Texas, where it has, till recently, been but little molested, it is still by no means a shy bird. . . .

"These birds feed on beechnuts, acorns (especially those of

the white and chinquapin oaks), chestnuts, pecan-nuts, black
persimmons, tuñas (the fruit of the prickly-pear), leguminous
seeds of various kinds, all the cultivated grains, different wild
berries and grapes, and the tender tops of plants; also grasshoppers, crickets, and other insects. The actions of the
gobbler during the mating-season, while paying court to the
female, are similar to those of the Domestic Turkey, and well
enough known to need no description.   .   .   .

"The call-notes of the Wild Turkey resemble those of the
domesticated bird very much; still they differ somewhat. In
feeding, the usual note is 'quitt, quitt,' or 'pit, pit.' When
calling each other it is 'keow, keow, kee, kee keow, keow,'
and a note uttered when alarmed suddenly sounds somewhat
like 'cut—cut.'" (*Bendire.*)

Nest.—A slight depression in the ground, either at the foot
of a tree or under a thick bush, and more or less lined with
dead leaves and grass.

Eggs.—Vary in number from eight to thirteen, but ten is
probably the general number. Occasionally two hens lay in
the same nest. Mr. G. E. Beyer writes:—"On May 25th,
1888, I found a nest with twenty-six eggs; one hen sitting
on the nest, and one standing by. I think both hens kept the
same nest."

SUB-SP. *a.* THE FLORIDA TURKEY. MELEAGRIS OSCEOLA.
Meleagris gallopavo osceola, Scott, Auk, 1890, p. 376.
Meleagris osceola, Ogilvie-Grant, Cat. B. Brit. Mus. xxii. p. 390
(1893).

Adult Male.—Differs from the typical *M. americana* in having
the white bands on the quills *very* narrow, much narrower
than the dark interspaces, and the tips of the tail-feathers paler
chestnut.

Range.—Florida, United States of America.

Habits.—Dr. William L. Ralph, of Utica, New York, writes:—
'Fifteen years ago I found the Wild Turkey abundant in

most parts of Florida, north of Lake Okeechobee, with perhaps the exception of the Indian River region, but they have
gradually decreased in numbers since then, and though still
common in places where the country is wild and unsettled,
they are rapidly disappearing from those parts, in the vicinity
of villages and navigable waters.

"One can hardly believe that the Wild Turkeys of to-day
are of the same species as those of fifteen or twenty years ago.
Then they were rather stupid birds, which it did not require much
skill to shoot, but now I do not know of a game-bird or mammal
more alert or more difficult to approach. Formerly, I have often,
as they were sitting in trees on the banks of some stream, passed
very near them, both in row-boats and in steamers, without
causing them to fly, and I once, with a party of friends, ran a
small steamer within twenty yards of a flock, which did not
take wing until several shots had been fired at them. . . .
These birds, though resident, are given to wandering a great
deal, and do not, like the Bob Whites, become attached to any
particular locality. At times they will remain in a favourable
place for weeks, but they are very uncertain, and will often
leave such a spot for no apparent reason. When they are
molested, or when there is a scarcity of food, they will keep in
motion most of the time during the day, and will often travel
many miles in a few hours.

"Wild Turkeys usually go in flocks, consisting of from two
or three to fifteen or twenty birds, and are also occasionally
found singly. Small flocks and single birds are more apt to
be found now than formerly, and the large droves, consisting
of several flocks associating together, are seldom if ever to be
seen of late. Their favourite places of resort are woods with
swamps in them or in their vicinity, and they always go to
these swamps to roost or when molested.

"These birds are polygamous, and the female takes all the
cares and duties of incubation upon herself. The gobblers
are very pugnacious, and will often fight fiercely for the favours

of the hens. The love-season begins in Florida about the middle of February and lasts for about three months, and during this period the gobblers frequently utter their call and are then easily decoyed within gun-shot. Native hunters have informed me that the hens roost by themselves at this season of the year."

Nest and Eggs.—Similar to those of *M. americana* described above.

III. THE HONDURAS TURKEY. MELEAGRIS OCELLATA.

Meleagris ocellata, Cuv. Mém. Mus. d'Hist. Nat. vi. p. 1, pl. i. (1820); Temm. Pl. Col. v. pl. 16 [No. 112] (1824); Elliot, Monogr. Phasian. i. pl. 33 (1872); Ogilvie-Grant, Cat. B. Brit. Mus. xxii. p. 391 (1893).

Meleagris aureus, Vieillot, Tabl. Encycl. Méth. i. p. 361 (1823).

(*Plate XXXI.*)

Adult Male.—Feathers of mantle, scapulars, chest, and flanks *brass-green,* shading into purplish-black towards the extremity, which is margined by a deep black line and fringed with greenish-copper; ends of upper tail-coverts and *ocelli on tail-feathers* greenish-blue, shot with purple; tail-feathers *widely margined with rich reddish-copper* changing to green; breast and belly black, margined with copper-red. Naked skin of head and neck *blue,* scarlet round the eye, and ornamented with *red* warts, the largest being situated between the eyes; erectile fleshy process on forehead *blue.* Tarsus armed with a long, stout, sharp spur. Total length, about 33 inches; wing, 14·2; tail, 13·1; tarsus, 4·4.

Adult Female.—Like the male, but less brilliantly coloured, and the occlli at the ends of the tail-feathers much reduced; the erectile process very small, and the great spurs represented by small wart-like knobs. Total length, about 33 inches; wing, 14·2; tail, 13·1; tarsus, 4·4.

Range.—Central America; Guatemala, Yucatan, and British Honduras.

PLATE XXXI

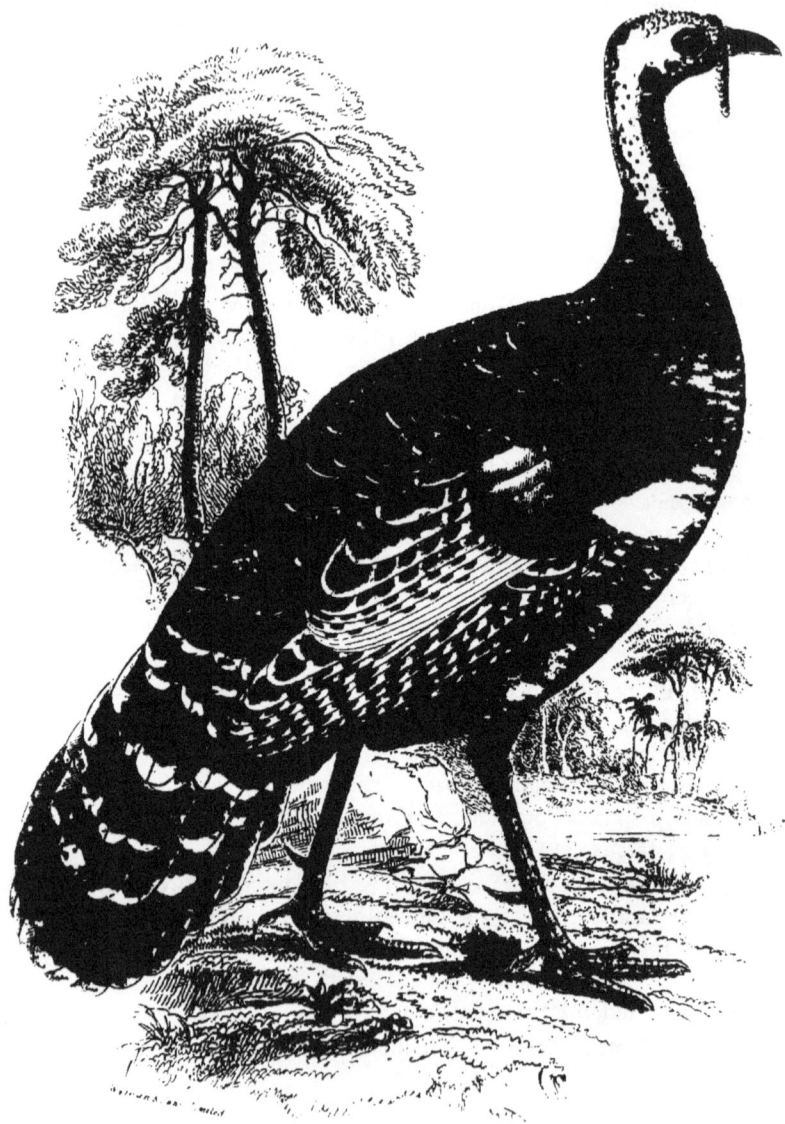

HONDURAS TURKEY.

Habits.—The only notes on the habits of this fine species are the following, furnished by Mr. F. Gaumer:—"The Spanish name for this bird is *Pavo del Monte*. It is occasionally seen within five leagues of Merida, but cannot be said to be common west of Espita. East of Espita it is often seen in the corn-fields in small flocks of from six to ten. I have discovered a locality ten leagues to the north and east of Valladolid, where it may be said to be common. This is the region depopulated since the emigration of the Indians nearly half a century ago; no one lives there now, and the *Meleagris* is the proud ruler of the forest. It is one of the wildest and shyest of birds, extremely cautious in its movements, and ever on the alert for a hidden enemy; it flies with the greatest rapidity at the sight of man, regardless of distance. When met with in open land it takes flight, rising with a heavy flutter peculiar to the family, and after mounting a few yards sails away with set wings to such a distance that the hunter never cares to follow. During the breeding-season, which is in May and June, the male makes a peculiar drumming noise, very deep and sonorous; after this he utters his peculiar song, which resembles the rapid pecking of a distant Woodpecker or the song of the great Bull Toad. On discovering a dreaded object, he utters a peculiar cluck and glides away with a proud movement, which seems to defy the world; and if the object moves, he darts away with headlong speed. The natives believe that this bird sees the image of its enemies in its plumage even before they are visible to the eye of the bird. However this may be, it is a bird of extraordinary caution and vision. Its flesh is held in the highest esteem by the natives, who hunt it unceasingly on this account. In Merida a specimen sells at from $1 to $2 dressed; and from $8 to $10 when alive. It is not easily domesticated, and rarely lives more than a few months."

Eggs.—Perfectly similar to those of the Common Turkey, but rather smaller. Measurements, 2·4 by 1·8 inches.

THE AMERICAN PARTRIDGES AND QUAILS.
SUB-FAMILY ODONTOPHORINÆ.

The following genera of American Partridges and Quails are distinguished from all the *Phasianidæ* previously described by having the cutting edge of the lower mandible serrated, and together form the Sub-family *Odontophorinæ* (see vol. i. p. 78). They vary considerably in size, some, such as *Dendrortyx*, being as large as the Common Partridge or slightly larger, while others are smaller than the Common Quail. In some instances the tooth-like process on the lower mandible is less distinct, but in the great majority of species it is easily recognised.

THE LONG-TAILED AMERICAN PARTRIDGES. GENUS DENDRORTYX.

Dendrortyx, Gould, Monogr. Odontoph. pl. 20 in pt. i. (1844); Introd. p. 20 (1850).

Type, *D. macrurus* (J. & S.).

Plumage alike in both sexes.

A short crest; a naked space round the eye.

First primary flight-feather *much shorter* than the tenth; fifth longest.

Tail as long, or nearly as long as the wing, wedge-shaped, composed of *twelve* feathers, the middle pair being *much longer* than the outer pair.

Tarsus *shorter* than the middle toe and claw.

Bill very stout and short.

Three rather large species about the size of the common Grey Partridge are known.

1. THE LONG-TAILED PARTRIDGE. DENDRORTYX MACRORUS.

Ortyx macroura, Jard. and Selb. Ill. Orn. i. text to pl. 38 and pl. 49 (1825-39).

Tetrao marmorata, La Llave, Reg. Trim. i. p. 144 (1831) ; id. La Nat. Mex. vii. App. p. 65 (1884).

Dendrortyx macrourus, Gould, Monogr. Odontoph. pt. i. pl. 2c (1844) ; Ogilvie-Grant, Cat. B. Brit. Mus. xxii. p. 392 (1893).

Adult Male.—Chin and throat *black* (feathers of the chin sometimes with whitish centres) ; short eyebrow-stripes and band bordering sides of throat *white ;* neck, mantle, chest, and sides chestnut broadly edged with grey ; lower back and rump dull olive-brown finely mottled with black ; wing-coverts and shoulder-feathers mostly olive-grey, with irregular black and white spots and markings ; breast and middle of belly dirty-white or brownish-olive ; naked space round the eye coral-red. Total length, 15 inches; wing, 6·3; tail, 6·5 ; tarsus, 2; middle toe and claw, 2·1.

Adult Female.—Smaller. Total length, 13·3 inches ; wing, 5·9; tail, 5·3; tarsus, 1·75 ; middle toe and claw, 1·8.

Range.—Southern Mexico; the highlands of Colima, Guerrero, and Oaxaca.

II. THE BEARDED LONG-TAILED PARTRIDGE.   DENDRORTYX BARBATUS.

Dendrortyx barbatus, Licht.; Gould, Monogr. Odontoph. pt. ii. pl. 22 (1846) ; Ogilvie-Grant, Cat. B. Brit. Mus. xxii. p. 393 (1893).

Adult.—Easily distinguished from *D. macrourus* by having the chin and throat *grey ;* eyebrow-stripes and cheeks dark grey ; chest, breast, and sides of belly light chestnut. Naked skin round the eye red. Total length, 13·5 inches ; wing, 6·2; tail, 4·8 ; tarsus, 1·95 ; middle toe and claw, 2·1.

Range.—South-east Mexico; the vicinity of Jalapa, Vera Cruz.

III. THE WHITE-EYEBROWED LONG-TAILED PARTRIDGE.

DENDRORTYX LEUCOPHRYS.

Ortyx leucophrys, Gould, P. Z. S. 1843, p. 132.

Dendrortyx leucophrys, Gould, Monogr. Odontoph. pt. ii. pl. 21 (1846); Ogilvie-Grant, Cat. B. Brit. Mus. xxii. p. 394 (1893).

Adult.—Forehead, eyebrow-stripe, chin, and upper-part of throat *white ;* feathers on lower part of throat, neck, and mantle grey, with chestnut centres edged on either side with black ; breast and under-parts dusky-grey, with a well-marked rufous stripe down the middle of each feather. Naked skin round the eye orange-red. Total length, 13·6 inches ; wing, 6 ; tail, 5·7 ; tarsus, 2·1.

Range.—Highlands of Guatemala and Costa Rica.

Habits.—Dr. A. Von Frantzius informs us that this beautiful Wood Partridge is called "Chirascuá" on account of its peculiar cry, which is heard before sunset. It is frequently met with in the neighbourhood of thick forests, and is much sought after on account of its savoury flesh, but, being very wild, is difficult to shoot. A specimen which was kept for a long while in a cage remained shy and wild to the last. It is often met with in the Poas, Candelaria, and Dota Mountains.

Eggs.—Oval and Grouse-like, reddish-buff, spotted and dotted with reddish-brown. Measurements, 1·75 by 1·25 inch.

THE SCALY PARTRIDGES. GENUS CALLIPEPLA.

Callipepla, Wagler, Isis, 1832, pp. 277, 1229.

Type, *C. squamata* (Vigors).

Sexes almost similar in plumage. Crest short, not extending much beyond the feathers of the head.

First primary flight-feather about equal in length to the eighth ; fourth longest.

Tail about three-fourths of the length of the wing, and composed of fourteen feathers.

Tarsus shorter than the middle toe and claw.

Only two small forms are known.

I. THE SCALY PARTRIDGE.   CALLIPEPLA SQUAMATA.

Ortyx squamatus, Vigors, Zool. Journ. v. p. 275 (1830).
Tetrao cristata, La Llave, Reg. Trim. i. p. 144 (1831); id. Nat. Mex. vii. App. p. 65 (1884).
Callipepla strenua, Wagler, Isis, 1832, p. 278.
Callipepla squamata, Gould, Monogr. Odontoph. pt. i. pl. 19 (1844); Cassin, Ill. B. Calif. p. 129, pl. xix. (1853); Bendire, N. Am. B. p. 18, pl. i. figs. 4, 5 [eggs] (1892); Ogilvie-Grant, Cat. B. Brit. Mus. xxii. p. 395 (1893).
Callipepla squamata pallida, Brewster, Bull. Nutt. Orn. Club, vi. p. 72 (1881).

Adult Male.—General colour above grey, browner on the wings and lower back, each feather of the neck, mantle, chest, and sides of breast edged with black, giving these parts a scaled appearance; throat and middle of breast and belly, whitish-buff; *no dark chestnut patch in the middle of the belly.* Total length, 10 inches; wing, 4·8; tail, 3·6; tarsus, 1·2; middle toe and claw, 1·35.

Adult Female.—Differs slightly in having dusky shaft-stripes to the feathers of the throat, and in being slightly smaller than the male.

Range.—New Mexico, Arizona, Western Texas, and North and Central Mexico, from Chihuahua and San Luis Potosi, as far south as the city of Mexico.

Habits.—"The Scaled Partridge, usually called the Blue Quail and also the White Top-knot Quail, is a constant resident in South-western Texas from about latitude 28° northward along the valley of the Rio Grande, as well as in a considerable por-

I 2

tion of New Mexico and Southern Arizona, extending south
into Mexico. . . .

" According to my own observations, the Scaled Partridge is
found most abundantly on the high plateaus bordering on the
principal streams of the regions under consideration, reaching
an altitude of from 1,500 to nearly 7,000 feet. It shuns timbered country, and in Southern Arizona, where I have frequently
met with these birds, they seemed to me to prefer the most
barren and dryest portions of that scantily-watered territory.
I invariably found them back in the foot-hills and mesas, from
two to five miles distant from the river beds, which are generally
dry for the greater part of the year. . . .

" From the repelling nature of the country it generally
frequents, it is naturally hunted very little; still, I found it
exceedingly shy and wary, and very difficult to approach, far
more so than Gambel's Partridge. It prefers to trust almost entirely to its legs for escape, and is generally successful, being an
expert and swift runner, dodging in and out among the bushes
with great ease and dexterity, and is consequently soon lost to
sight. The covey generally follow a leader, Indian-file fashion.
Its habits seem, however, to differ somewhat in other parts
of its range." (*Bendire.*)

Mr. E. W. Nelson furnishes the following observations
about this species. He says:—" In many instances I have
found them far from water, but they make regular visits to the
watering-places. . . . They are very difficult to flush,
owing to the rapidity with which they run through the bushes
and other vegetation. When flushed, they scatter, and after
flying a short distance, alight, and run on as before. As soon
as the alarm is over, the old birds reassemble the flock by a
low call-note.

" In the latter part of summer and early fall they gather into
coveys, often containing several broods, as I observed in 1882,
in the valley of the Gila River, near Clifton, Arizona. At this
season they frequented the low bare hill-sides or the now dry

water-courses and the fields adjoining these, associating with Gambel's Partridge. They are easily trapped in the fall and winter, and many are caught by the natives and taken to the markets of the larger towns of New Mexico and Arizona."

Mr. William Lloyd writes from Marfa, Texas, as follows:— "The Blue Quails love a sandy table-land, where they spend considerable time in taking sand-baths. I have often watched them doing so, pecking and chasing each other like a brood of young Chickens. Good clear water is a necessity to them. They are local, but travel at least three miles for water. In the evenings they retire to the smaller ridges or hillocks, and their calls are heard on all sides as the scattered covey collects. Several times I have seen packs numbering from sixty to eighty, but coveys from twenty-five to thirty are much oftener noticed."

Nest.—Placed on the ground under the shelter of a small bush, in corn- and grain-fields, in meadows, potato-fields, or almost barren flats. (*Bendire.*)

Eggs.—Vary from nine to sixteen, but eleven or twelve are generally found. Pale creamy-white to pale buff, finely dotted or spotted all over with reddish-brown, vinaceous-buff, or fawn-colour in different sets. Average measurement, 1·3 by 1 inch.

SUB-SP. *a*. THE CHESTNUT-BELLIED SCALY PARTRIDGE.
CALLIPEPLA CASTANEIVENTER.

Callipepla squamata castanogastris, Brewster, Bull. Nutt. Orn. Club, viii. p. 34 (1883); Bendire, N. Am. B. p. 22, pl. i. figs. 6, 7 [eggs] (1892).

Callipepla castaneiventer, Ogilvie-Grant, Cat. B. Brit. Mus. xxii. p. 396 (1893).

Adult Male.—Differs from the typical *C. squamata* in having the middle of the breast and belly deeper buff or ochreous ; *a dark reddish-chestnut patch on the middle of the belly.*

Adult Female.—The chestnut patch on the middle of the belly absent or rudimentary.

Range.—Tamaulipas, North-eastern Mexico, and the lower Rio Grande Valley, Texas.

"The general habits of the Chestnut-bellied Scaled Partridge, as well as its food, are very similar to those of the preceding sub-species. The mating- and nesting-season, however, commences somewhat earlier." (*Bendire.*)

Nest and Eggs.—Similar to those of *C. squamata.*

THE PLUMED PARTRIDGES. GENUS OREORTYX.

Oreortyx, Baird, B. N. Amer. p. 642 (1860).

Type, *O. pictus* (Dougl).

Sexes almost similar in plumage. A crest of *two very long* feathers.

First primary flight-feather *intermediate in length* between the seventh and eighth; third or fourth slightly the longest.

Tail composed of *twelve* feathers, and about *three-fifths* of the length of the wing.

Tarsus *shorter* than the middle toe and claw.

Only one species, rather larger than the Common Quail, is known.

I. THE PLUMED PARTRIDGE OR MOUNTAIN QUAIL.
OREORTYX PICTUS.

Ortyx picta, Douglas, Trans. Linn. Soc. xvi. p. 143 (1823); Jard. and Selb. Ill. Orn. ii. pl. 107.

Ortyx plumifera, Gould, P. Z. S. 1837, p. 42 ; Aud. B. Amer. v. p. 69, pl. 291 (1842).

Oreortyx pictus, Baird, B. N. Amer. p. 642 (1860); Bendire, N. Amer. B. p. 13 (1892); Ogilvie-Grant, Cat. B. Brit. Mus. xxii. p. 397 (1893).

Oreortyx pictus plumiferus, Bendire, N. Am. B. p. 14, pl. i. figs. 2, 3 [eggs] (1892).

Adult Male.—*Crest of two very long black feathers ;* head, *neck, mantle*, and breast *slate-grey ;* rest of upper-parts mostly olive-brown; throat and fore-neck deep chestnut margined by a white band; a black patch on the cheek; belly chestnut, shading into pale buff; flanks barred with chestnut, white, and black. Total length, 9·6 inches; wing, 5·2 ; tail, 3·3 ; tarsus, 1·4 ; middle toe and claw, 1·6.

Adult Female.—Differs as a rule in having the *olive-brown continued up the back of the neck to the crest*, but in some examples the upper mantle is more or less washed with grey.

Range.—Western States of North America ; extending northwards to Washington Territory, southwards through California to Cape St. Lucas, and westward to Eastern Nevada.

Habits.—Mr. Charles A. Allen, writing to Captain Bendire, says :—" I find this Partridge all through the Sierras. In the spring many go up to the snow-line, returning in the fall below the point of snowfall. These vertical migrations are performed *entirely on foot*, unless streams must be crossed, when they take to their wings, but alight at once on gaining the opposite side, and continue their travels on foot."

Captain Bendire writes :—" The mating season begins in the latter part of March and the beginning of April, according to latitude and altitude. The call-note of the male is a clear whistle, like ' whu-ié-whu-ié,' usually uttered from an old stump, the top of a rock, or a bush. When alarmed, a note like ' quit-quit ' is used. In the higher mountains but a single brood is raised ; but in the lower foot-hills they rear two broods occasionally, the male caring for the first one while the female is busy hatching the second.

"I met with a brood of young birds, perhaps a week or ten days old, near Jacksonville, Oregon, on June 17, 1883. The male, in whose charge they were, performed the usual tactics of feigning lameness, and tried his very best to draw my attention away from the young, uttering meantime a shrill sound resembling ' quaih-quaih,' and showed a great deal of distress,

seeing I paid no attention to him. The young, already handsome and active little creatures, scattered promptly in all directions, and the majority were most effectually hidden in an
instant. As nearly as I was able to judge they numbered
eleven. I caught one, but after examining it turned it loose
again. The feathers of the crest already showed very plainly.

" Their food consists of insects, the buds and tender tops of
leguminous plants, small seeds, and berries of various kinds."

Nest.—Generally on the ground among a collection of dead
leaves, and well concealed by bushes or ferns, while a favourite
site is said to be beneath the fallen tops of pine trees left by
wood-cutters. Occasionally nests are met with on the tops of
old decayed tree-stumps.

Eggs.—Generally from eight to twelve in number, but sometimes as many as sixteen may be found. The colour varies
from pale cream to reddish-buff, and the shell is devoid of
markings. Average measurements, 1·38 by 1·06 inch.

THE CALIFORNIAN QUAILS. GENUS LOPHORTYX.

Lophortyx, Bonap. Comp. List B. Eur. and N. Amer. pp. 42,
        43 (1838).

        Type, *L. californicus* (Shaw and Nodder).

Plumage of sexes different. Crest in both sexes extending
much beyond the feathers of the head.

First primary flight-feather *intermediate in length* between
the eighth and ninth ; fourth generally slightly longer than the
third and fifth.

Tail composed of *twelve* feathers, and about *three-fourths* of
the length of the wing.

Tarsus *shorter* than the middle toe and claw.

Three species about the size of the Common Quail are
known.

I. THE CALIFORNIAN QUAIL. LOPHORTYX CALIFORNICUS.

Tetrao californicus, Shaw and Nodder, Nat. Misc. ix. p. 345 (1797 ?).

Ortyx californicus, Jardine and Selby, Ill. Orn. i. pl. 38, ii. pl. 107 ; Aud. B. Amer. v. p. 67, pl. 290 (1842).

Callipepla californica, Gould, Monogr. Odontoph. pt. i. pl. 16 (1844) ; Bendire, N. Am. B. p. 23, pl. i. figs 8–10 [eggs] (1892).

Lophortyx californicus, Bonap. ; Ogilvie-Grant, Cat. B. Brit. Mus. xxii. p. 400 (1893).

Lophortyx californicus brunnescens, Ridgw. P. Biol. Soc. Wash. ii. p. 94 (1884).

Callipepla californica vallicola, Ridgw. P. U. S. Nat. Mus. viii. p. 355 (1886) ; Bendire, N. Am. B. p. 26 (1892).

Adult Male.—Crest of *black* club-shaped feathers ; *throat* and cheeks *black*, margined by a white band ; eyebrow-stripes and a band between the eyes white ; sides and back of neck grey, margined with black and spotted with white ; mantle, chest, and tail grey ; lower back and rump greyish olive-brown ; wing rather darker ; middle of breast buff, shading into chestnut on the belly, and *both margined with black ; sides olive-grey.* Total length, 9·7 inches ; wing, 4·4 ; tail, 3·6 ; tarsus, 1·25 ; middle toe and claw, 1·45.

Adult Female.—Crest *shorter and browner; no* black and white pattern on head ; *throat white, with dark shaft-stripes ;* neck, mantle, and chest *brownish-grey ;* under-parts *white,* tinged with buff on the belly ; *sides and flanks olive-brown.* Slightly smaller.

Range.—Western States of North America, extending north to Washington, south as far as Cape St. Lucas, Lower California, and eastwards to Nevada. It has been introduced into various parts of the world.

Habits.—"Their favourite haunts," says Prof. O. B. Johnson, "are the undergrowth and thickets along water-courses, brush

covered side hills, and cañons, and they frequent the roads,
cultivated fields, vineyards, and edges of clearings to feed.
It is a constant resident, and breeds wherever found.

"The mating time commences early in March, sometimes
later, depending on the season. Then the large packs into
which this species gathers in the fall of the year, break up
gradually, each pair of birds selecting a suitable nesting-site.
In the more densely settled portions of California, this Partridge
is by no means as common now as it was a decade ago, when
it was not unusual to see packs numbering five hundred and
more together, while now, at least near the larger cities, coveys
even of fifty birds are rarely seen. In localities where it is
not constantly harassed and hunted, the Californian Partridge
becomes surprisingly tame and confiding, in fact almost domesticated; and under such circumstances many nest close to
houses and out-buildings and in the shrubbery of gardens adjoining human habitations.

"The young run about as soon as they are hatched. Usually
but one brood is raised, occasionally two. In the latter case
the male takes charge of the young when they are about three
weeks old, the female then laying the eggs for the second.
Downy young have been observed as early as the 20th of May
in the southern portions of their range, and some broods are
undoubtedly hatched still earlier. In the fall, when the young
are full grown and able to shift for themselves, they collect
in large packs, a number of coveys associating together until
the spring. They are much shyer then, and more difficult to
approach. The usual call-note, when one of these packs becomes scattered, is a rather unmusical 'ca-āpe, ca-āpe,' the
last syllable drawn out; another note, like 'kā-kāāh,' is also
used on such occasions."

Mr. A. W. Anthony writes about the Lower California birds
as follows:—"I found the Valley Partridge (Californian Quail)
very common in the mountains of Lower California, up to an
altitude of about 9,000 feet. Both in Southern and Lower

California I was told by the Indians and native Mexicans that during very dry seasons the Valley Quail did not nest, but remained in large flocks during the entire summer. This statement I was able to verify by personal observations during the summer of 1887. These birds were seen by me in large flocks throughout the spring and summer months, and only two or three broods of young were noticed. Birds taken during April, May, and June showed but little development of the ovaries. Should the winter rains, however, be sufficient to insure an abundance of seeds and grasses, the coveys begin to break up early in March, and from every hill in the land the loud challenge of the male is heard. . . . By far the most common call at all seasons is one resembling 'ca-ra-ho,' repeated four or five times, and the accent shifted from one syllable to another as suits the fancy of the performer."

Mr. William Proud, writing from Butte County, California, informs Captain Bendire as follows :—" In early seasons they begin to pair in the last week of February, but the time varies somewhat according to the season. During this period there is considerable fighting among the males for the favour of the coveted female. This is kept up until all are suitably mated and the nesting-season arrives. This usually begins here about the last week in March, when the pairs scatter among the shrubbery along the banks of creeks and in adjacent ravines, along hedgerows and brush fences, and on the borders of cultivated fields. The earliest nest I ever found was on March 15th, and on April 15th I met young birds probably a couple of days old. I consider fourteen eggs to be about the average number laid by these birds, and have found as many as twenty-four in a nest. The large sets I attribute to other hens laying in the nest, probably young birds which have failed to make preparation for their own eggs. On May 21 my dog pointed a Valley Partridge on a nest which contained twenty-two eggs, and every one hatched.

" During incubation the male is very attentive and watchful, usually taking an elevated position near the nest, where, with crest erect, and tail spread, he bids defiance to all intruders, uttering an oft-repeated 'whew-whew-whew.' When the brooding hen leaves the nest to feed, should he be absent from the post of duty, her cry of 'tobacco, tobacco,' very plainly given, brings him up at once. In fact their call-notes are very varied. I frequently heard an old cock call out at night 'ah-hooh, ah-hooh,' the first note in a low key."

Nest.—Generally a slight hollow scratched in the ground and slightly lined. The situation varies greatly. The shelter afforded by a rock, log, or old stump, small bush, bunch of weeds or grass, is usually selected, but occasionally a perfectly open situation is chosen, and even trees are sometimes resorted to, the fork formed by two large branches, or the upright end of a broken decayed limb being chosen as a site.

Eggs.—Generally from twelve to sixteen in number, but sometimes many more are found, probably the result of two hens laying in the same nest. Ground-colour creamy-white, sometimes buff, spotted and dotted, or blotched all over with reddish-brown or olive-drab. Average measurements, 1·28 by 1 inch.

II. GAMBEL'S QUAIL.   LOPHORTYX GAMBELI.

Lophortyx gambeli, Nutt. ; Gambel, P. Ac. Philad. 1843, p. 260 ; Ogilvie-Grant, Cat. B. Brit. Mus. xxii. p. 403 (1893).
Callipepla venusta, Gould, P. Z. S. 1846, p. 70.
Callipepla gambeli, Gould, Monogr. Odontoph. pt. iii. pl. 17 (1850) ; Bendire, N. Am. B. p. 29, pl. i. figs. 11–14 [eggs] (1892).

Adult Male.—Easily distinguished from the male of *L. californicus* by having the back of the head *chestnut; no white* spots on the back of the neck ; the chest-feathers with dark shafts ; the middle of the belly *black ; no* black margins to the feathers of the breast and belly, and the sides *deep chestnut,* with white centres. Total length, 9·8 inches; wings, 5; tail, 4; tarsus, 1·3; middle toe and claw, 1·55.

Adult Female.—Differs from the female of *L. californicus* in having the crest considerably more developed ; *no* white spots on the back of the neck ; the chest-feathers with dark shafts ; *no* black margins to the feathers of the breast and belly, and the flanks *chestnut* with white centres. Rather smaller in size.

Range.—Western States of North America ; extending north to Utah, south to the State of Sonora, North Mexico, westward to South California, and east to Arizona and Western Texas.

Habits.—From Captain Bendire's splendid work, " Life Histories of North American Birds," the following notes on this species are taken :—" Wherever water is found, Gambel's Partridge is common throughout Southern Arizona up to an altitude of 5,000 feet ; and in New Mexico Mr. W. H. Cobb, of Albuquerque, informs me that he met with young fledglings in the pine forests at an altitude of 8,000 to 9,000 feet. . . . During the mating- and breeding-season, the former commencing usually in the latter part of February, the latter about the first week in April, and occasionally later, according to the season, the male frequently utters a call like ' yuk-käe-ja, yuk-käe-ja,' each syllable distinctly articulated and the last two somewhat drawn out. A trim, handsome, and proud - looking cock, whose more sombre-coloured mate had a nest close by, used an old mesquite stump, about four feet high, and not more than twenty feet from my tent, as his favourite perch, and I had many excellent opportunities to watch him closely. Standing perfectly erect, with his beak straight up in the air, his tail slightly spread, and wings somewhat drooping, he uttered this call in a clear strong voice every few minutes for half an hour or so, or until disturbed by something, and this he repeated several times a day. I consider it a call of challenge, or of exultation, and it was taken up usually by any other male in the vicinity at the time. During the mating-season the males fight with each other persistently, and the victor defends his chosen home against intrusion with much valour. . . .

"During the intense heat of the Arizona summers, Gambel's Quail, like most other birds, prefers to remain in the shady and cool spots in the creek bottoms, frequently perching in the trees, and I believe the majority of these birds spend the nights in them as well. They take to trees very readily at all times."

Nest.—A slight hollow scratched in the ground, usually lined with bits of dry grass and sheltered by dead grass or old brush ; sometimes placed among grain-fields. Occasionally nests are found in situations above the ground, the top of a rotten stump or an old nest of some other species being made use of.

Eggs.—Generally from ten to twelve in number, but much larger numbers are sometimes found, no doubt the product of more than one hen. Ground-colour creamy-white or pale-buff, spotted, clouded, or blotched with reddish-brown or dark brown. Average measurements, 1·26 by ·96

III. DOUGLAS'S QUAIL.   LOPHORTYX DOUGLASI.

Ortyx douglasii, Vigors ; Douglas, Trans. Linn. Soc. xvi. p. 145 (1828).

Ortyx elegans, Lesson, Cent. Zool. p. 189, pl. 61 (1830).

Ortyx spilogaster, Vigors, P. Z. S. 1832, p. 4.

Callipepla elegans, Gould, Monogr. Odontoph. pt. i. pl. 18 (1844).

Lophortyx douglasi, Ogilvie-Grant, Cat. B. Brit. Mus. xxii. p. 404 (1893).

Callipepla elegans bensoni, Ridgw. P. U. S. Nat. Mus. x. p. 148 (1887).

Adult Male.—Crest long and *pale rufous ;* throat black, *each feather edged with white ;* feathers of the back of the head, neck, and upper mantle with a *triangular chestnut spot* at the end of each feather ; inner wing-coverts, shoulder-feathers, sides, and flanks *similarly marked*, but with a white spot or partial margin on each web ; breast and belly with *round white spots.* Total length, 9·3 inches ; wing, 4·4 ; tail, 3·2 ; tarsus, 1·2 ; middle toe and claw, 1·4.

Adult Female.—Crest generally *dark brown ;* throat *white, with dark shaft-stripes ;* under-parts greyish-brown with round white spots.

Range.—Western Mexico ; States of Sonora, Sinaloa, and Jalisco.

THE BARRED PARTRIDGES. GENUS PHILORTYX.

Philortyx, Gould, Monogr. Odontoph. pl. 14 in pt. ii. (1846); Introd. p. 17 (1850).

Type, *P. fasciatus* (Gould).

Sexes *similar* in plumage. A *well-developed crest* extending much beyond the feathers of the head.

First primary flight-feather *intermediate in length* between the ninth and tenth ; fourth slightly the longest.

Tail composed of *twelve* feathers, nearly *three-fourths* of the length of the wing.

Tarsus shorter than the middle toe and claw.

Only one small species is known.

I. THE MEXICAN BARRED PARTRIDGE. PHILORTYX FASCIATUS

Ortyx fasciatus, Natterer, MS. ; Gould, P. Z. S. 1843, p. 133

Ortyx perrotiana, Des Murs, Rev. Zool. 1845, p. 207.

Philortyx fasciatus, Gould, Monogr. Odontoph. pt. ii. pl. 14 (1846); Ogilvie-Grant, Cat. B. Brit. Mus. xxii. p. 406 (1893).

Philortyx personatus, Ridgw. Auk, iii. p. 333 (1886).

Adult.—Crest blackish-brown tipped with rufous ; neck and mantle olive-grey, shading into olive-brown on the back, which

is mixed *with black and margined with buff;* wings and shoulder
feathers very similar, but the black markings take the form of
blotches near the tips ; throat white ; under-parts white, *strongly
barred with black,* except the middle of the breast and belly.
Total length, 7·6 inches ; wing, 4 ; tail, 2·6 ; tarsus, 1·15 ;
middle toe and claw, 1·35.

Younger Birds have the eyebrow-stripes and greater part of the
chin and throat *black.*

Range.—Southern Mexico. States of Colima, Guerrero, and
Puebla.

THE CRESTED QUAILS. GENUS EUPSYCHORTYX.

Eupsychortyx, Gould, Monogr. Odontoph. pl. 10 in pt. i.
   (1844) ; Introd. p. 15 (1850).

Type, *E. leucopogon* (Lesson).

Sexes *different* in plumage. Crest well or fairly *well-developed.*

First primary flight-feather *intermediate in length* between
the eighth and ninth ; fourth slightly the longest.

Tail composed of *twelve* feathers, rather *more than half* the
length of the wing.

Tarsus shorter than the middle toe and claw.

Eight small Quail-like species are known.

I. THE CURAÇAO CRESTED QUAIL. EUPSYCHORTYX CRISTATUS.

Tetrao cristatus, Linn. S. N. i. p. 277 (1766).

Ortyx temmincki, Steph. in Shaw's Gen. Zool. xi. p. 381
   (1819).

Ortyx neoxenus, Vigors, P. Z. S. 1830, p. 3.

Eupsychortyx cristatus, Gould, Monogr. Odontoph. pt. ii. pl.
   9 (1846) ; Hartert, Ibis, 1893, pp. 305, 325 ; Ogilvie-
   Grant, Cat. B. Brit. Mus. xxii. p. 407 (1893).

Perdix neoxenus, Aud. Orn. Biogr. v. p. 228, pl. 423 (1849).

Eupsychortyx gouldi, Berlepsch, J. f. O. 1892, p. 100.

Adult Male.—*Crest*, top of the head, and *throat, buff or rufous-buff ; the broad eyebrow-stripes* and bands bordering the sides of the throat *black;* ear-coverts white ; back of neck white, spotted with black ; back and wings mostly rufous-grey, shading into grey on the rump, and all finely mottled and blotched with black, and the former edged with buff; chest, sides, and flank-feathers rufous down the middle barred with black and spotted with white ; rest of under-parts white barred with black, and widely edged with orange-buff towards the middle. Total length, 8 inches ; wing, 4 ; tail, 2·4 ; tarsus, 1·15 ; middle toe and claw, 1·3.

Adult Female.—Differs chiefly from the male in having the *eyebrow-stripes orange-yellow*, the ear-coverts *pale-brown*, and the feathers on the sides of the throat whitish-buff margined with black on the sides.

Range.—The islands of Curaçao and Aruba.

It was not till 1892, when Mr. E. Hartert visited the islands of Curaçao and Aruba, which lie off the northern coast of Venezuela, that the true home of this species became known. Though the bird was accurately described by Brisson in 1760, the locality whence it came had ever been a matter of doubt, and though Gould, in his monograph of the *Odontophorinæ*, gives the habitat as "Mexico," it has never been found there by recent explorers. It is possible that it may occur in Venezuela.

Habits.—Mr. Hartert tells us that "this pretty bird is not rare on Aruba and Curaçao, but is not found everywhere. The natives call it 'Socklé,' a name derived from its note, which is uttered very frequently. It is much esteemed as food, and sometimes sold in the market alive.

"This bird is not easy to obtain in any great numbers without a dog, as it does not care to fly, and is difficult to be seen in grassy places. It is not found on Bonaire," which lies immediately to the east of Curaçao.

II. THE WHITE-FACED CRESTED QUAIL.   EUPSYCHORTYX
LEUCOPOGON.

Ortyx leucopogon, Lesson, Rev. Zool. 1842, p. 175; Des Murs,
Icon. Orn. pl. 36 (1846) [crest omitted].
Ortyx leucotis, Gould, P. Z. S. 1843, p. 133.
Eupsychortyx leucotis, Gould, Monogr. Odontoph. pt. i. pl. 10
(1844).
Eupsychortyx leucopogon, Gould, Monogr. Odontoph. pt. iii.
pl. 13 (1850); Ogilvie-Grant, Cat. B. Brit. Mus. xxii. p.
409 (1893).

Adult Male.—Very similar to the male of *E. cristatus,* but
the crest is brownish-white; eyebrow-stripes *rufous-chestnut,*
and the *white* ear-patches margined above and below with
chestnut instead of black. From the following species it differs
in having the *chest thickly spotted* right up to the throat.
Total length, 8·5 inches; wing, 4·1; tail, 2·7; tarsus, 1·2;
middle toe and claw, 1·3.

Adult Female.—Differs from the male in having the crest *dark
brown;* the feathers of the eyebrow-stripe and throat buff,
edged with black; the ear-coverts brown; the mantle blotched
with black like the rest of the back; the under-parts paler and
the cross-bars more strongly marked; somewhat smaller.

Range.—Veragua to the United States of Colombia.

III. SONNINI'S CRESTED QUAIL.   EUPSYCHORTYX SONNINII.

Perdix sonnini, Temm. Pig. et Gall. iii. pp. 451, 737 (1815); id.
Pl. Col. v. pl. 42 [No. 75] (1823).
Eupsychortyx sonnini, Gould, Monogr. Odontoph. pt. iii. pl. 11
(1850); Ogilvie-Grant, Cat. B. Brit. Mus. xxii. p. 410
(1893).

Adult Male.—Differs from the male of *E. leucopogon* in having
the ear-coverts dirty white; the chest almost *uniform pale vinaceous,* slightly mottled with black (there are irregular white black-

edged spots on one or both margins of the chest-feathers, but these are mostly hidden). Total length, 8·5 inches; wing, 4·1 ; tail, 2·6 ; tarsus, 1·15 ; middle toe and claw, 1·3.

Adult Female.—Much like the female of *E. leucopogon*, but with the eyebrow-stripes and cheeks marked with *orange-rufous, and much brighter ;* the general colour of the under-parts *paler,* and the black markings *less coarse.*

Range.—North of South America, extending south to the Rio Branco, east to British Guiana, and west to Caracas, Venezuela. Introduced into St. Thomas, West Indies.

IV. THE CUMANÁ CRESTED QUAIL. EUPSYCHORTYX MOCQUERYSI.

Eupsychortyx mocquerysi, Hartert, Bull. B. O. C. no. iii. p. 37 (1894); id. Ibis, 1894, p. 430; id. Novit. Zool. i. pl. xv. fig. 2.

Adult.—Differs from *E. sonninii* in the following points :— The throat is white all along the middle, most of the feathers showing distinct narrow cross-bars of black. *The breast is uniform vinaceous-cinnamon,* except on the lower part ; the belly and sides of the body are similarly coloured, but varied with large white spots bordered with black. Total length, 9 inches ; wing, 4·1–4·25; tail, 2·6 ; tarsus, 1·1 ; middle toe and claw, 1·35.

Range.—Cumaná, on the north coast of Venezuela.

V. THE SHORT-CRESTED QUAIL. EUPSYCHORTYX PARVICRISTATUS.

Ortyx parvicristatus, Gould, P. Z. S. 1843, p. 106.
Eupsychortyx parvicristatus, Gould, Monogr. Odontoph. pt. ii. pl. 12 (1846); Ogilvie-Grant, Cat. B. Brit. Mus. xxii. p. 410 (1893).

Adult Male.—Like the male of *E. sonninii* but the crest is short and dark brown ; the ear-coverts dark brown ; the chest pale chestnut; the breast and belly rufous-chestnut, with fewer white spots. Total length, 8·5 inches; wing, 3·9 , tail, 2·5; tarsus, 1; middle toe and claw, 1·15.

Adult Female.—Similar to the female of *E. sonninii.*

Range.—United States of Colombia.

VI. LEYLAND'S CRESTED QUAIL.   EUPSYCHORTYX LEYLANDI.

Ortyx leylandi, Moore, P. Z. S. 1859, p. 62.
Eupsychortyx leucofrenatus, Elliot, Ann. Lyc. N. York, vii. p. 106, pl. 3 (1860).
Eupsychortyx leylandi, Ogilvie-Grant, Cat. B. Brit. Mus. xxii. p. 411 (1893).

Adult Male.—Chin and throat *black;* as in the two last-mentioned species, the chest vinaceous, indistinctly mottled with dusky, and with a few white spots; mantle *dark grey,* coarsely mottled and marked with black. Total length, 8·4 inches; wing, 4·1; tail, 2·4; tarsus, 1·1; middle toe and claw, 1·3.

Adult Female.—Closely resembles the females of the last two species, but the *chest* is like that of the male, only more spotted; the eyebrow-stripes broad and *pure yellowish-buff;* the throat *buff,* the outer feathers slightly edged with black, and the upper-parts more coarsely marked with black.

Range.—Central America; Honduras, Nicaragua, and Costa Rica.

Habits.—M. Boucard met with this species in the valley of San José, Costa Rica, and found it common in the coffee-plantations during the rainy season, from May to December, but rare in the other months, when it disappeared completely from the valley. He also met with small coveys in the plains, and observes that they can run extremely fast.

Mr. G. C. Taylor frequently saw coveys of these birds in Honduras, especially on the high ground near Comayagua. They were usually lying in long grass, and, when disturbed, used to fly for shelter into the thick bushes. They were difficult to raise without a dog, and very difficult to see when on the wing. Moreover, the ground they frequented was so full of ticks and "garrapatas," as to destroy all keenness in the pursuit of them. In habits this species appears to resemble the common Virginian Quail (*O. virginianus*).

This beautiful Partridge, called in Costa Rica, "Perdiz," is often found over the whole plateau in flocks of from fifteen to twenty, as well in the open country in the neighbourhood of thick underwood, as in the wheat-fields surrounding the Heredia and Barbee. (*Dr. A. von Frantzius.*)

VII. THE BLACK-THROATED CRESTED QUAIL. EUPSYCHORTYX NIGROGULARIS.

Ortyx nigrogularis, Gould, P. Z. S. 1842, p. 181 ; id. Monogr.
    Odontoph, pt. ii. pl. 4 (1846) ; G. R. Gray, Gen. B. iii.
    p. 514, pl. cxxxii. (1846).
Colinus nigrogularis segoviensis, Ridgw. P. U. S. Nat. Mus. x. p.
    593 (1887).
Eupsychortyx nigrogularis, Ogilvie-Grant, Cat. B. Brit. Mus.
    xxii. p. 412 (1893).

Adult Male.—Crest moderately developed, *brown ;* eyebrow-stripes, chin, and throat *black ;* a *white stripe* between the angle of the gape and the ear-coverts ; mantle *chestnut,* most of the feathers with a fairly distinct *white* central spot ; chest and under-parts *white,* each feather margined with *black,* except the sides and flanks, which are edged with chestnut. Total length, 8·2 inches ; wing, 4·1 ; tail, 2·3 ; tarsus, 1·15 ; middle toe and claw, 1·35.

Adult Female.—Distinguished from the female of *E. sonninii* and the allied species by having the eyebrow-stripes and throat *bright buff, without any trace* of black markings.

Range.—Central America ; Yucatan, British Honduras, and Honduras.

Habits.—In Yucatan the Black-throated Quail was always seen by Mr. G. F. Gaumer in flocks or pairs, sometimes in the darkest forests, but more usually in corn-fields. Unfortunately, he tells us nothing further respecting its habits, merely remarking that the flesh is delicious. According to Dyson, it frequents the pine ridges, and is very common both in Honduras and Yucatan, moving about in coveys. From Dr. Samuel Cabot's notes we take the following, which give a more detailed account: —"In reading works relating to the discovery and conquest of Yucatan by the Spaniards, we see mention made of the sacrifices of Quails offered by the natives to their idols; sometimes the blood only was offered, and sometimes the whole body. The bird there alluded to is undoubtedly the *Ortyx nigrogularis*, as this is the only bird called 'Codorniz' or 'Quail' by the Spanish residents of the country. The *Ortyx nigrogularis* in its note and habits is precisely similar to the *O. virginianus*. They whistle the 'Bob-White' in spring; their covey-call in the autumn and winter is so precisely the same that they readily answered when I whistled the call of our Quail; and if I had previously scattered the covey, I could always find them in this way. They feed on similar food, and roost in the same way; they also sometimes alight on trees, like our Quail. . . . The Maya or Indian name of this bird is Bêch, the *e* pronounced with a guttural sound."

VIII. THE WHITE-BREASTED CRESTED QUAIL. EUPSYCHORTYX HYPOLEUCUS.

Eupsychortyx hypoleucus, Gould, P. Z. S. 1860, p. 62 ; Ogilvie-Grant, Cat. B. Brit. Mus. xxii. p. 413 (1893).

Adult Male.—Easily distinguished from all the other species of this genus by having the eyebrow-stripes, chin, throat, and under-parts pure *white*. Total length, 7·8 inches ; wing, 4·1 ; tail, 2·5 ; tarsus, 1·1 ; middle toe and claw, 1·3.

Adult Female.—Closely resembles the female of *E. cristatus*, but the feathers of the crest are darker.

Range.—Central America ; Guatemala.

THE COLINS OR BOB-WHITES. GENUS ORTYX.

Ortyx, Steph. in Shaw's Gen. Zool. xi. p. 376 (1819).
Type, *O. virginianus* (Linn.).
Sexes different in plumage. *No* distinct crest.
First primary flight-feathers *intermediate in length* between the seventh and eighth ; fourth slightly the longest.
Tail composed of twelve feathers, rather *more than half* the length of the wing.
Tarsus shorter than the middle toe and claw.
Ten small Quail-like forms are known.

I. THE VIRGINIAN COLIN OR BOB-WHITE. ORTYX VIRGINIANUS.

Tetrao virginianus, T. marilandicus, and *T. mexicanus,* Linn.
S. N. i p. 277 (1766).
Perdix virginiana, Wilson, Am. Orn. vi. p. 21, pl. xlvii.
(1812); Aud. Orn. Biogr. i. p. 388, pl. 76 (1831), v. p.
564 (1839).
Ortyx virginianus, Aud. B. Amer. v p. 59, pl. 289 (1842);
Gould, Monogr. Odontoph. pt. i. pl. 1 (1844); Ogilvie-
Grant, Cat. B. Brit. Mus. xxii. p. 415 (1893).
Colinus virginianus, Bendire, N. Am. B. p. i. pl. i. fig. i. [egg].
(1892.)

Adult Male.—*Chin and throat white, surrounded by a black band ;* a *black band* from the gape to the chestnut ear-coverts ; feathers of the mantle *vinaceous-rufous,* the *edges* grey, barred with black ; middle of the breast and belly white or whitish-buff, with the V-shaped black bars narrower and less marked. Total length, 8·5 inches ; wing, 4·5 , tail, 2·5 ; tarsus, 1·2 ; middle toe and claw, 1·45.

Adult Female.—Distinguished from the male by having the throat *bright buff,* the black bands from the gape to the ear-coverts and round the throat ill-defined, and the black bars on the under-parts nearly obsolete, especially on the middle of the breast and belly.

Range.—Eastern United States of North America, extending north to Massachusetts and Minnesota, west to Dakota and Indian Territory, and south to Georgia and other Gulf States. Introduced into many of the West Indian Islands and various parts of the Old World.

Habits.—"This species, one of the most widely distributed of our Game-Birds, is better known throughout the Northern and Middle States as the Quail, and under the name of Partridge, or Virginia Partridge, in the South. . . .

"At the present time the Bob-Whites are most abundant in the Central and some of the Southern States. They have also been successfully introduced in various localities in the West. . . .

"Excepting, perhaps, in its extreme northern range, the Bob-Whites are residents, and breed wherever found. They are partial to more or less open country. Fields and pastures, interspersed with small bodies of woodland, country roads, bordered by brush and briar patches, as well as the edges of meadow and lowlands, are their favourite abiding places. In Southern Louisiana they are very partial to the borders of hammock land and open pine-woods. They are never found in large packs; each covey generally keeps to itself, and rarely moves far from the place where it was raised. The mating-season commences in April, when the coveys, or such portions of them remaining, begin to break up, each pair selecting a suitable nesting-site. Nidification begins usually about May; in the Southern States somewhat earlier, and in the more northern portions of their breeding-range it is often delayed until June. . . .

"These birds are very sociable in disposition, and, when not constantly disturbed or shot at, become quite tame, and may frequently be seen about dwelling-houses, barns, and in gardens, especially during the late fall, winter, and early spring. As soon as the young are hatched, they become more shy and retiring. The young leave the nest as soon as hatched, and have been seen running about with pieces of the shell sticking to them. They are faithfully cared for by both parents, who make use of all sorts of artifices, such as feigning lameness and fluttering along just out of reach of the intruder, to lure him away from the young brood ; the young scattering in the meantime, and hiding in the grass and under leaves at the danger-signal of the parents, and remaining quiet until called together again by either of them as soon as all danger is passed. When they are about two or three weeks old, the male takes charge of the first brood, while the female begins to lay her second clutch of eggs. This is usually a smaller one than the first, averaging only about twelve eggs. The young are at first exclusively fed on insect food, and later on small seeds, grains, and berries. . . . The males commence singing about the 1st of May; their song is the well-known 'Bob-White,' or 'Ah, Bob-White.' One of their love-notes may be translated as ' Pease most ripe,' another call as ' No more wet,' or ' More wet.' A shrill ' Wee-tech ' is used as a note of warning, and one to assemble when the covey has dispersed, resembles 'Quoi-hee, quoi-hee.' A subdued clucking when undisturbed, and a rapidly-repeated twitter when suddenly surprised, are frequently used as well." (*Bendire.*)

Nest.—A cavity scratched in the ground and sheltered by overhanging weeds, grass, or brush, &c. " Occasionally the nest is arched over, but in most cases, where there is no natural cover existing, no dome is attempted." (*Bendire.*)

Eggs.—This bird is the most prolific of North American Game-Birds, the number of eggs varying from twelve to eighteen. " As many as thirty-seven eggs have been found in one nest, unquestionably the product of two, or even three, hens. In such large sets the eggs are always placed in layers

or tiers, the small or pointed ends always towards the centre."
(*Bendire.*)

Dull white, slightly glossed, often partially stained with
yellowish-buff. Average measurements, 1·2 by ·96 inch.

SUB-SP. *a.* THE FLORIDA COLIN. ORTYX FLORIDANUS.

Ortyx virginianus floridanus, Coues, Key N. Amer. B. p. 237
    (1872); Ogilvie-Grant, Cat. B. Brit. Mus. xxii. p. 418
    (1893).

Colinus virginianus floridanus, Bendire, N. Amer. B. p. 7
    (1892).

Adult Male.—Differs from the male of *O. virginianus* in
having the general tone of the plumage darker; ear-coverts
black, the band from the gape passing *uninterruptedly* across
them round the base of the throat, where it *widens out, often
extending over the upper chest;* the black bars on the under-
parts much coarser and more strongly marked. Measurements
the same.

Adult Female.—Darker than the female of *O. virginianus*, and
with the black bars on the under-parts much more strongly
marked, and *equally defined on the middle* of the breast and
belly.

Range.—Florida. Cuba; probably introduced.

Habits.—Captain Bendire writes :—"This somewhat smaller
and darker race is found only in Florida. Dr. W. L. Ralph,
who has enjoyed excellent opportunities for studying the habits
of the Florida Bob-White, and is well-known as a reliable and
careful observer, writes to me as follows :—' It is still common
throughout the northern and central parts of the State, and
probably in the southern portions as well, but they are not
nearly so abundant as formerly, owing to the persecution they
receive from northern visitors and negroes, and to the want of
efficient game laws. They are very tame and confiding, and
when not molested prefer to live near man, probably on

account of greater security from the attacks of beasts and
birds of prey. They become much attached to the localities
where they breed, and seldom wander far from these, even
when much persecuted. I have known cases where they were
hunted day after day until their number was reduced to two
or three birds to each covey, yet those which were left could
always be found at their old places of resort. The localities
they like best are open woods grown up with sand palmettes
or low bushes, or fields with woods near them, and they are
particularly fond of slovenly cultivated grounds that have
bushes and weeds growing thickly along their borders.'"

Nest and Eggs.—Similar to those of *O. virginianus.*

SUB-SP. *b.* THE TEXAN COLIN. ORTYX TEXANUS.

Ortyx texanus, Lawrence, Ann. Lyc. N. York, vi. p. i. (1853);
Baird, B. N. Amer. p. 641, pl. xxiv. (1860); Ogilvie-
Grant, Cat. B. Brit. Mus. xxii. p. 419 (1893).
Colinus virginianus texanus, Bendire, N. Am. B. p. 8 (1892).

Adult Male.—Differs from *O. virginianus* and *O. floridanus*
in having the feathers of the mantle barred and mottled with
pale rufous and black, and indistinctly edged with grey ; as in
the former, the black band round the base of the white throat
is *narrower,** but, like the latter, the black bars on the under-
parts are coarser and more marked. Measurements the same.

Adult Female.—Like the female of *O. floridanus,* but the
general colour of the upper-parts is *greyer and paler.*

Range.—Southern and Western Texas, and North-east and
Western Mexico.

Habits.—The Texan Bob-White is a resident in the greater
part of Texas, excepting the so-called Staked Plains in the
north-western part of the State. In Eastern Texas it intergrades

* There are two males in the British Museum collection with a large
black patch covering the chin and middle of the throat. These are
apparently mere individual varieties.

with *Ortyx virginianus*. It is most abundant in the central parts of the State. Its range northward extends well into the Indian territory, and it has also been taken in Western Kansas, where, however, it is rare. Its general habits do not differ materially from those of *O. virginianus*. (*Bendire.*)

Mr. William Lloyd, of Marfa, Texas, says:—" The Texan Bob-Whites are birds of the lowlands, and not found above an altitude of 2,000 feet. Their food consists of small berries, acorns, grain, buds and leaves of aromatic herbs and small shrubs, varied with occasional beetles, grasshoppers, and ants, especially the winged females, of which they seem to be very fond. They are very unsuspicious, and their low notes, uttered while feeding, attract a good many enemies. I have seen foxes on the watch, and the Marsh Harrier perched in a clump of grass on the look-out, waiting for them to pass. But the many large Rattlesnakes found here are their worst enemies. One killed in May had swallowed five of these birds at one meal ; another a female, evidently caught on her nest, and a half-dozen of her eggs ; a third, four Bob-Whites and a Scaled Partridge."

Nest and Eggs.—Similar to those of *O. virginianus*.

II. THE CUBAN COLIN.   ORTYX CUBANENSIS.

Ortyx cubanensis, Gould, Monogr. Odontoph. pt. iii. pl. 2 (1850); Ogilvie-Grant, Cat. B. Brit. Mus. xxii. p. 421 (1893).

Colinus virginianus cubanensis, Bendire, N. Am. B. p. 9 (1892).

Adult Male.—Chin and throat white, surrounded by a black band ; top of the head *black ;* chest-feathers black, mixed with dull rufous or white in the middle ; rest of under-parts chestnut, irregularly edged with black, and spotted with white on the sides. Total length, 8 inches ; wing, 4·2 ; tail, 2·3 ; tarsus, 1·2 ; middle toe and claw, 1·4.

Adult Female.—Coarsely spotted on the under-parts with black, dirty white, and rufous.

Range.—Island of Cuba ; Isle of Pines.   It is also reported from Porto Rico.*

Habits.— Dr. Jean Gundlach, who is intimately acquainted with the habits of this species, says that the " Codorniz " is only met with towards the outskirts of the forest, and inhabits the extensive prairies in the most westerly part of the Island of Cuba.   The flight of the bird is neither high nor protracted, but straight and swift ; it rises with rapid beats of the wing, which become less laboured when the bird is well on the wing, or are sometimes entirely suspended, the wings being merely widely extended.   When the flight is caused by approaching danger, the different birds of the flock or covey separate from each other, and settle some way off in different places, some-times running for a little distance.   When the danger has dis-appeared, all reassemble at the call of the leader of the flock. If the panic has been caused by dogs, the birds fly on to the nearest tree, where they crouch on a horizontal branch, and remain motionless as long as the dog stays under the tree or barks.   They may then, if approached with caution, be caught by means of a loop made of horse-hair or strong thread fastened to a long slender pole.   If the flock takes flight without being scared by any danger, the different members all fly together.

They search for their food on the ground, picking up various kinds of seeds, and sometimes berries or young leaves. Should they anticipate any danger whilst thus employed, a murmuring sound is heard, and they run with raised crest, outstretched necks, and outspread tails to a place of safety.   Their call-notes vary somewhat, according to circumstances.   In the mating-season, when the members of a flock are already paired or scattered about, the cock perches on a branch, stump of a tree, post, stone, or a large clod of earth, and summons his mate with two or three notes, the third being quickly or loudly uttered, and the hen answers.   The cocks often fight with each

* It is possible that the following notes by Dr. J. Gundlach may refer to the Florida Colin, which has apparently been introduced into Cuba.

other. Their flesh is good, and, considering the size of the birds, there is a wonderful amount of meat on them. The male bird takes part in hatching the eggs, and should the first brood fail, a second set of eggs is laid. As in all species of this order, the newly-hatched young run about as soon as they are dry.

The "Codorniz" is caught in traps, and can be easily kept in a cage, but when in captivity their feathers in time become very rough. Dr. Gundlach had no experience of their nesting in cages or aviaries, but had seen a hen take to a newly-hatched chicken and rear it.

Nest.—Built between the middle of April and July. A hollow in the ground lined with a few dry grasses, &c., and sheltered by projecting plants.

Eggs.—Ten to eighteen in number; white. Measurements, 1·2 by 1 inch.

III. THE BLACK-BREASTED COLIN.   ORTYX PECTORALIS.

Ortyx pectoralis, Gould, P. Z. S. 1841, p. 182 ; id. Monogr. Odontoph. pt. iii. pl. 5. (1850); Ogilvie-Grant, Cat. B. Brit. Mus. xxii. p. 421 (1893).

Adult Male.—Throat white, surrounded by a black band which *extends over the upper part of the chest;* upper-parts much like those of *O. texanus ; under-parts uniform pale rufous-chestnut.* Total length, 7·7 inches; wing, 4 ; tail, 2·1; tarsus, 1·1 ; middle toe and claw, 1·25.

Adult Female.—Most like the female of *O. texanus,* but the upper-parts are *darker and browner*, and the *black markings* on the under-parts *heavier.* Measurements as in the male.

Range.—Vera Cruz, Eastern Mexico.

IV. GRAYSON'S COLIN.   ORTYX GRAYSONI.

Ortyx graysoni, Lawrence, Ann. Lyc. N. York, viii. p. 476 (1867); Ogilvie-Grant, Cat. B. Brit. Mus. xxii. p. 422 (1893).

(*Plate XXXII.*)

PLATE XXXII.

GRAYSON'S COLIN.

Adult Male.—Differs chiefly from the male of *O. pectoralis* in having the black band round the base of the throat narrow, not extending over the chest, which is dull rufous-chestnut like the rest of the under-parts. Total length, 7·8 inches; wing, 4·5; tail, 2·5; tarsus, 1·2; middle toe and claw, 1·4.

Adult Female.—Like the female of *O. pectoralis*, but slightly larger.

Range.—State of Jalisco, West Mexico.

V. RIDGWAY'S COLIN.   ORTYX RIDGWAYI.

Colinus ridgwayi, Brewster, Auk, ii. p. 199 (1885); Bendire, N. Amer. B. p. 10 (1892).

Ortyx ridgwayi, Ogilvie-Grant, Cat. B. Brit. Mus. xxii. p. 422 (1893).

Adult Male.—Distinguished from all the preceding species by having the eyebrow-stripes, *chin, and throat black;* and from the two following species by having the *chest* and under-parts pale reddish brick-colour. Total length, 8·4 inches; wing, 4·5; tail, 2·5; tarsus, 1·2; middle toe and claw, 1·35.

Adult Female.—Closely resembles the females of *O. pectoralis* and *O. graysoni.* Measurements as in the male.

Range.—Arizona, United States of North America, and Sonora in Northern Mexico.

Habits.—Mr. Herbert Brown, who was the first to obtain examples of this species, gives the following note:—"The habits of the Masked Bob-White, so far as we know them, appear to resemble very closely those of the Common Quail (*O. virginianus*), only slightly modified by the conditions of their environment. They utter the characteristic call of 'Bob-White' with bold, full notes, and perch on rocks or bushes while calling. They do not appear to be mountain-birds, but live on the mesas (table-lands) in the valleys and possibly in the foot-hills.

"In addition to their ' Bob-White ' they have a second call of ' Hoo-we,' articulated and as clean cut as their ' Bob-White.' This call of ' Hoo-we ' they use when scattered, and more especially when separated towards nightfall. At this hour I noted that, although they occasionally called 'Bob-White,' they never repeated the first syllable, as in the daytime they now and then attempted to do."

Nest and Eggs.—Similar to those of *O. virginianus*. Average measurements of the pure white eggs, 1·25 by 1 inch.

VI. THE COYOLCOS COLIN. ORTYX COYOLCOS.

Tetrao coyolcos, P. L. S. Müll. S. N. Suppl. p. 129 (1776).
Ortyx coyolcos, Gould, Monogr. Odontoph. pt. iii. pl. 6 right-
    hand fig. (1850); Ogilvie-Grant, Cat. B. Brit. Mus. xxii.
    p. 423 (1893).

Adult Male.—Like the males of *O. pectoralis* and *O. graysoni*, and, as in the last species, the chin and throat are black, but the *chest is black ;* the feathers of the top of the head are black, *edged with brown* ; and the eyebrow-stripes, if present, are *white*, and indistinctly represented. Total length, 7·7 ; wing, 4·2 ; tail, 2·3 ; tarsus, 1·15 ; middle toe and claw, 1·4.

Adult Female.—Like that of *O. ridgwayi*, but smaller. Measurements as in the male.

Range.—Oaxaca, Southern Mexico.

VII. THE BLACK-HEADED COLIN. ORTYX ATRICEPS.

Ortyx coyolcos, Gould (*nec* Müll.), Monogr. Odontoph. pt. iii.
    pl. 6 left-hand fig. (1850).
Ortyx atriceps, Ogilvie-Grant, Cat. B. Brit. Mus. xxii. p. 424
    (1893).

(*Plate XXXIII.*)

Adult Male.—Like the male of *O. coyolcos*, but the top of the head, eyebrow-stripe, chin, and throat are all *uniform black*,

PLATE XXXIII.

BLACK-HEADED COLIN.

Wyman & Sons Lmitd

and the general colour of both upper- and under-parts is darker. Measurements as in *O. coyolcos.*

Adult Female.—Differs from the female of *O. coyolcos* in being *altogether darker*, the grey markings of the mantle being replaced by brownish-black. Total length, 7·4 inches; wing, 4; tail, 2·3; tarsus, 1·1; middle toe and claw, 1·35.

Range.—Putla, Western Mexico.

VIII. THE CHESTNUT-COLOURED COLIN. ORTYX CASTANEUS.

Ortyx castaneus, Gould, P. Z. S. 1842, p. 182, id. Monogr. Odontoph. pt. iii. pl. 3 (1850); Ogilvie-Grant, Cat. B. Brit. Mus. xxii. p. 424 (1893).

Adult Male.—Top of the head, mantle, chest, and general colour of the rest of the upper-parts *dark chestnut*, the chin and throat *black;* middle of the breast and belly white, barred with black and mixed with chestnut. Total length, 9 inches; wing, 4; tail, 2·3; tarsus, 1·1; middle toe and claw, 1·35.

But a single example of this bird is known—Gould's type preserved in the British Museum Collection. The locality and other particulars are wanting. It has been suggested by some American ornithologists that the specimen in question is merely a strongly-marked variety of *O. virginianus,* and it may possibly transpire that it is so. In support of this theory we may remind our readers that the so-called Mountain Partridge (*see* vol. i. p. 147, pl. xii.) is undoubtedly nothing but a strongly-marked rufous variety of the Common Partridge.

THE HARLEQUIN QUAILS. GENUS CYRTONYX.

Cyrtonyx, Gould, Monogr. Odontoph. pl. 7 in pt. i. (1844), Introd. p. 14 (1850).

Type, *C. montezumæ* (Vig.).

Sexes differ in plumage. A rather full crest, but none of the feathers very elongate.

First primary flight-feather *intermediate in length* between the seventh and eighth ; fourth slightly longest.

Tail composed of *twelve* feathers, *less than half* the length of the wing.

Tarsus shorter than the middle toe and claw.

Claws very long.

I. THE MASSENA HARLEQUIN QUAIL.   CYRTONYX MONTEZUMÆ.

Ortyx montezumæ, Vigors, Zool. Journ. v. p. 275 (1830) ; Jardine and Selby, Ill. Orn. iii. pl. 126.

Ortyx massena, Lesson, Ill. Zool. pl. 52 (1831).

Tetrao guttata, La Llave, Reg. Trim. i. p. 114 (1831).

Odontophorus meleagris, Wagler, Isis, 1832, p. 277.

Cyrtonyx massena, Gould, Monogr. Odontoph. pt. i. pl. 7 (1844) ; Cassin, Ill. B. Cal. p. 21, pl. iv. (1853).

Cyrtonyx montezumæ, Bendire, N. Am. B. p. 35, pl. i. fig. 15 [egg] (1892) ; Ogilvie-Grant, Cat. B. Brit. Mus. xxii. p. 425 (1893).

Adult Male.—Crest reddish-buff; forehead, sides of the head, throat, and fore-neck elegantly marked with a *black-and-white pattern* ; general colour above pale rufous, thickly barred with black, with wide buff shaft-stripes ; outer webs of quills barred with white ; middle of chest and breast dark chestnut; sides and flank *dark grey*, with a row of rounded *paired white spots* down each web ; rest of under-parts deep black. Total length, 8·2 ; wing, 5·1 ; tail, 2·2 ; tarsus, 1·1 ; middle toe and claw, 1·4.

Adult Female.—Head and throat *without* black-and-white pattern, and mostly white, the eyebrow-stripes, and cheeks washed with pale vinaceous ; base of fore-neck bordered by an ill-defined black band, commencing on the ear-coverts ; upper-parts more coarsely marked than in the male ; under-parts pale vinaceous, shading into buff towards the middle, and more or less spotted and marked with black. Total length, 7·8 inches ; wing, 4·7 ; tail, 2·1 ; tarsus, 1·1 ; middle toe and claw, 1·35.

Range.—Mexico, extending north to Arizona and South-western Texas, south as far as the city of Mexico, west to Jalisco, and east to Tamaulipas.

Habits.—Mr. Dresser procured several examples of this species, also known as the "Fool Quail" or "Black Part-ridge," in Southern Texas, and remarks :—"I afterwards found the bird on several occasions when riding along the higher hill ranges, and altogether shot six, while a man who was with me killed two more. In their habits they are more like the Texan Quail than any other, but on the wing are easily dis-tinguished, for they fly heavily, though very swiftly. When dis-turbed they squat very close, and will not move until one is close upon them ; indeed, I found them generally rise up almost under my feet."

Mr. William Lloyd, writing to Captain Bendire from Marfa, Texas, says that "the favourite resorts of the Massena Part-ridge are the rocky ravines or arroyas that head well up in the mountains. They quickly, however, adapt themselves to changed conditions of life, and are now to be seen around the ranches picking up grain and scratching in the fields. In the vicinity of Fort Davis, Texas, they have been exceptionally numerous, and may frequently be seen sitting on the stone walls surrounding grain-fields in Limpa Cañon. In Mexico I have seen them several times living contentedly in cages. In Mes-quite Cañon they are the only Partridge found ; and in June and July, 1887, I spent some time there trying principally to locate the nest and eggs of this species. I found a single egg in a depression at the roots of a tasaca cactus, presumably belonging to this species. It was white, without any markings whatever. While there, I was informed by two different parties living in the vicinity that each of them had found a nest the previous year, 1886, containing eight and ten eggs respectively, which they had eaten. They described the eggs as being

L 2

white in colour. Both said that the nests were simply a slight
hollow, one under a small shin-oak bush, the other alongside
a sotol plant."

According to Mr. John Swinburne, of St. John's, Apache
County, Arizona, the favourite localities frequented by this
species during the breeding-season are thick live-oak scrub
and patches of rank grass, at an altitude of from 7,000 to
9,000 feet. He says :—" Here they are summer residents only,
descending to much lower altitudes in winter. They lie very
close at all times, allowing one to almost step on them before
they move. I have seen this species on the white mountains during the breeding-season, and saw young birds of the year shot
there. Even the adults seem very stupid when suddenly
flushed, and, after flying a short distance, alight, and attempt
to hide in most conspicuous places. I have seen men follow
and kill them by throwing stones."

Nest.—A hollow scratched in the ground under the shelter
of a long grass, &c., and generally more carefully lined than
that of *O. virginianus*.

Eggs.—Eight to ten in number, rather glossy, and somewhat
pointed towards the smaller end ; pure white when laid, but
often discoloured after the bird has commenced to sit.
Average measurements, 1·28 by ·9 inch.

II. SALLÉ'S HARLEQUIN QUAIL.   CYRTONYX SALLÆI.

Cyrtonyx sallæi, Verr. Arcana, Nat. i. p. 35, pl. 4 (1859) ;
    Ogilvie-Grant, Cat. B. Brit. Mus. xxii. p. 427 (1893).

Adult Male.—Easily distinguished from the male of *C.
montezumæ* by having the shaft-stripes on the upper-parts
(except those of the mantle) *orange-rufous* or *chestnut ;* the
stripe round the eye and triangular patch on the cheek dark
bluish-black ; the sides and flanks paler grey ; and the row
of spots on each web rufous-buff or chestnut.

Female.—Is not yet known.

Range.—Mexico ; State of Guerrero.

III. THE OCELLATED HARLEQUIN QUAIL. CYRTONYX
OCELLATUS.

Ortyx ocellatus, Gould, P. Z. S. 1836, p. 75.

Cyrtonyx ocellatus, Gould, Monogr. Odontoph. pt. ii. pl. 8
(1846); Ogilvie-Grant, Cat. B. Brit. Mus. xxii. p. 428
(1893).

Cyrtonyx sumichrasti, Lawrence, Ann. N. Y. Sci. i. p. 51
(1877).

Adult Male.—Differs from the *male* of *C. sallæi*, already described, in having the black markings on the upper-parts in the
form of *round black spots ;* the middle of the chest and breast
pale buff, tipped with rufous ; and the flanks chestnut, *irregularly barred with black*, shading into grey towards the margins.
Total length, 8·3 inches ; wing, 5·3 ; tail, 2·2 ; tarsus, 1·25 ;
middle toe and claw, 1·5.

Adult Female.—Like the female of *C. montezumæ*, but the
general colour above is *black*, finely barred with rufous and
mottled with sandy. Total length, 8 inches; wing, 5 ; tail
2·2 ; tarsus, 1·25 ; middle toe and claw, 1·45.

Range.—Central America ; Tehuantepec to Guatemala.

THE LONG-NAILED PARTRIDGES. GENUS
DACTYLORTYX.

Dactylortyx, Ogilvie-Grant, Cat. B. Brit. Mus. xxii. p. 429
(1893).

Type, *D. thoracicus* (Gambel).

Plumage of sexes *different.* A short crest.

First primary flight-feather *equal to* the eighth ; fourth
longest.

Tail composed of *twelve* feathers, and *two-fifths* of the
length of the wing.

Tarsus shorter than the middle toe and claw.

Claws *very long* and but *slightly curved.*

I. THE LONG-NAILED PARTRIDGE. DACTYLORTYX
THORACICUS.

Ortyx thoracicus, Gambel, P. Ac. Philad. iv. p. 77 (1848).
Odontophorus lineolatus, Gould, Monogr. Odontoph. pt. iii. pl.
    32 (1850).
Dactylortyx thoracicus, Ogilvie-Grant, Cat. B. Brit. Mus. xxii.
    p. 429 (1893).

Adult Male.—Crown, nape, and mantle reddish-brown, mixed
with black, and with pale shoulder-stripes ; shoulder-feathers
and wings very similar but largely mixed with black ; lower
back and rump mostly olive-brown ; eyebrow-stripes, cheeks,
and throat, *reddish-chestnut ;* a *black patch* on the sides of the
throat ; under-parts grey, tinged with reddish-brown, and with
white shafts ; middle of belly and vent white.   Total length,
9 inches ; wing, 5·3 ; tail, 2·2 ; tarsus, 1·3 ; middle toe and
claw, 1·7.

Adult Female.—Upper-parts much like those of the male ;
eyebrow-stripes and cheeks *greyish-white ;* throat *white ;* chest
and breast *dull brick-red*, with pale shafts.   Slightly smaller.

Range.—Central America ; Southern Mexico, Yucatan, Guatemala, and San Salvador.

Habits.—According to Mr. Salvin, "this is perhaps the
commonest Partridge found on the Volcan de Fuego.   The
ravines of this volcano are localities very favoured by several
species of the group.   It is not often, however, that they are
to be found actually at the bottom of the hollow, where the
increasing shadow and height of the overhanging trees render
the undergrowth of vegetation comparatively scanty, but most
frequently near the top of either side, in places where a fallen
tree or a slip of soil has laid bare a sunny spot.   Such situations
are sought for by these birds to bask and sleep in, like Partridges in a warm hedge-side.   They are, however, true forest-
birds, and are usually met with in small flocks of six or eight,
probably the brood of the season.

"When frightened, the whole bevy runs up the side of the ravine, and only when approached quite suddenly do they take wing. The consequence is (alas! that it should be said) that the sportsman is obliged to shoot them on the ground; and the only mode he has of quieting his conscience is, by a stretch of the imagination, to suppose them 'fur,' and not 'feather,' and take a running shot."

Mr. G. F. Gaumer says:—"This bird is common in all the eastern forests of Yucatan, where it is much esteemed for its fine flesh and as a household pet. As a pet it is not a success, living but a few months in confinement. Like the Quails, this bird lives upon the ground, where it is always seen in pairs. At nightfall it sings a very pretty song, beginning with a low whistle, which is three times repeated, each time with greater force; then follow the syllables *che-va-lieu-a* repeated from three to six times in succession. The tone is musical, half sad, half persuasive, beginning somewhat cheerfully, and ending more coaxingly. From its colour and its habit of remaining immovable while one is passing, this bird is somewhat difficult to see. I have frequently seen it squatting close to the ground while I passed within a few feet of it. It seldom flies, and never flies far when compelled to take wing."

THE THICK-BILLED PARTRIDGES. GENUS ODONTOPHORUS.

Odontophorus, Vieillot, Analyse, p. 51 (1816).

Type, *O. guianensis* (Gmel.).

Sexes similar or somewhat different in plumage. A moderately long and full crest.

First primary flight-feather *shorter than the tenth;* fourth or fifth rather the longest.

Tail composed of *twelve* feathers, *not more than half* the length of the wing.

Tarsus *not longer* than the middle toe and claw.

For convenience the fourteen forms may be divided into two groups :—

A. Chest and breast not spotted with white (species 1-8, pp. 152-158).

B. Chest and breast spotted with white (species 9-14, pp. 158-161).

A. Chest and breast not spotted with white.

I. THE GUIANA PARTRIDGE.   ODONTOPHORUS GUIANENSIS.

Tetrao guianensis, Gmel. S. N. i. pt. ii. p. 767 (1788).
Perdix dentata, Temm. Pig. et Gall. iii. pp. 419, 734 (1815).
Perdix rufina, Spix, Av. Sp. Nov. ii. p. 60, pl. 766 (1825).
Odontophorus rufus, Vieillot, Gal. Ois. ii. p. 38, pl. 211
    (1825).
Odontophorus guianensis, Gould, Monogr. Odontoph. pt. i. pl.
    25 (1844); Ogilvie-Grant, Cat. B. Brit. Mus. xxii. p. 432
    (1893).

Adult Male and Female.—Crown mostly deep chestnut ; nape and mantle grey, finely mottled with black ; lower back and rump reddish-brown, more or less dotted with black ; wings rufous, blotched and marked with black ; outer webs of primary quills *barred with buff ; cheeks, chin,* and sides of throat *dark chestnut ;* middle of throat grey ; general colour of chest and rest of under-parts *brownish-buff,* indistinctly barred with dusky. Naked skin round eye vermilion.

Adult Male.—Total length, 11·5 inches ; wing, 5·9 ; tail, 2·8 ; tarsus, 1·6 ; middle toe and claw, 1·8.

Adult Female.—Total length, 10 inches ; wing, 5·4 ; tail, 2·3 ; tarsus, 1·5 ; middle toe and claw, 1·6.

Habits.—This species frequents the forests, and is never seen in the savannas or open country. It is a shy bird, and usually met with singly or in pairs, never in coveys, and, like the rest of its kind, runs with great speed, but when flushed betakes itself to the branches of trees. It has two distinct notes; one a rather loud whistling call, which may be heard at morning and evening, the other a sound like "Tock'ro," whence its Macusi name. It is also known as the Duraquara.

Mr. T. K. Salmon writes:—" When wandering one morning in the forest I saw a pair engaged in the work of nest-making. The male was in the nest; and the female appeared to be building around him. The female made off at my approach, but the male continued in the nest until I nearly put my hand on him, no doubt trusting to his dark colour among the leaves to escape detection. I do not think I should have seen him had it not been for the scarlet over the eye."

Nest.— Builds its nest into a bank or side of the ground in the high forest, with a tunnel-like entrance made of interlaced twigs and sticks, or, perhaps, more properly speaking, with a neatly executed bow in front of the nest, which is merely a hole scraped in the ground and lined with dead leaves. (*Salmon.*)

Eggs.—Six to eight in number; white; measurements, 1·5 by 1·1 inch.

SUB-SP. *a.* THE MARBLED PARTRIDGE. ODONTOPHORUS
MARMORATUS.

Ortyx (Odontophorus) marmoratus, Gould, P. Z. S. 1843, p. 107.

Odontophorus pachyrhynchus, Tschudi, Fauna Peruana, p. 282 (1844–46); Gould, Monogr. Odontoph. pt. iii. pl. 24 (1850).

Odontophorus marmoratus, Ogilvie-Grant, Cat. B. Brit. Mus. xxii. p. 433 (1893).

Adult Male and Female.—Appear to be distinguished from *O. guianensis* by having the sides of the head and chin rust-red, and the general colour of the under-parts darker, and almost invariably barred with black and buff. The sexes do not appear to differ in size, and the measurements are rather larger

than in the male of *O. guianensis*. This form appears to inter-
grade with typical *O. guianensis*, and it seems doubtful if it is
even sub-specifically distinct, but the material at present available is insufficient to decide this question.

Range.—Western South America, ranging north into Panama
and southwards through the United States of Colombia, Ecuador, and Peru to Bolivia.

Tschudi met with this bird in the Andes of Peru, at an
elevation of from 4,000 to 7,000 feet, but he has published
nothing regarding its habits.

II. THE CAPUEIRA PARTRIDGE.   ODONTOPHORUS CAPUEIRA.

Perdix capueira, Spix, Av. Sp. Nov. ii. p. 59, pl. 76 A. (1825).
Ortyx capistrata, Jardine and Selby, Ill. Orn. i. text to pl. 38.
Odontophorus dentatus, Gould (*nec* Temm.), Monogr. Odontoph.
    pt. ii. pl. 26 (1846).
Odontophorus capueira, Ogilvie-Grant, Cat. B. Brit. Mus. xxii.
    p. 434 (1893).

Adult Male and Female.—Distinguished from *O. guianensis* by
having the mantle brown or rufous-brown, blotched with black,
and with pale buff shaft-stripes; the outer webs of the primary-
quills barred with *white;* and the chin, throat, and under-parts
dark grey. Total length, 11 inches; wing, 6·2 ; tail, 3·1 ; tarsus, 1·8; middle toe and claw, 1·95.

Range.—Eastern South America, extending north to Bahia,
west to Goyaz, and south to Rio Grande do Sul.

Habits.—According to Maximilian, Prince of Wied, this
bird is called Capueira by the Brazilians and closely resembles
the European Hazel Grouse (*Tetrastes bonasia*) in its habits and
mode of life.   It is never met with in the open country, keeping
entirely to the thick woods.   In the early part of the year the
Capueira is found in pairs, and after the breeding-season the
families remain in coveys of from ten to sixteen or more in

number. They are active birds, running very quickly and pro-
curing their food among the fallen leaves in the midst of the
vast forests. The stomachs examined contained fruits, berries,
insects, small stones, and a little sand. The loud and remarkable voice of this bird is only heard in the forests, where it commences the call before daybreak, the sound reverberating to a
great distance. Azara says that the cry is uttered by both
sexes, but the Prince was of opinion that only the male bird
called. During the morning and evening twilight the Capu-
eiras perch on a branch in a line very close one to another,
and at this time the male birds frequently give vent to their cry.
The sport afforded by these birds closely resembles that offered
by Hazel Grouse ; when a covey is disturbed by the dogs, they
fly into trees and may sometimes be easily killed, but at other
times they are lost sight of in the dense foliage.

Nest.—Placed on the ground.

Eggs.—Ten to fifteen in number ; pure white.

III. THE BLACK-EARED PARTRIDGE. ODONTOPHORUS MELANOTIS.

Odontophorus melanotis, Salvin, P. Z. S. 1864, p. 586 ; Ogilvie-
Grant, Cat. B. Brit. Mus. xxii. p. 435 (1893).

Adult Male.—Top of the head and eyebrow-stripes *uniform
chestnut ;* ear-coverts, chin, and throat *black ;* general colour of
the chest and under-parts *deep chestnut.* Total length, 9·5
inches ; wing, 5·8 ; tail, 2·2 ; tarsus, 1·75 ; middle toe and claw,
1·85.

Adult Female.—Probably differs from the male in having the
inner webs of the secondary quills tipped with buff, but I have
been unable to examine any female specimens in which the
sex has been actually ascertained.

Range.—Central America ; Nicaragua, Costa Rica, and
Veragua.

Habits.—Very few notes are to be found respecting the habits

of the Black-eared Partridge. Mr. C. W. Richmond writes
that he saw a flock of rather over a dozen in the forests on the
Escondido. When approached, the birds flew into the surrounding trees and afterwards off into the woods two or three
at a time. Two were secured.

IV. THE CHESTNUT-EARED PARTRIDGE. ODONTOPHORUS
ERYTHROPS.

Odontophorus erythrops, Gould, P. Z. S. 1859, p. 99; Ogilvie-
Grant, Cat. B. Brit. Mus. xxii. p. 435 (1893).

Adult Male.—Like *O. melanotis*, but distinguished by its darker
plumage; the top of the head *dark brown*, contrasting with
the rufous chestnut eyebrow-stripes; stripe from the gape to
the ear-coverts *chestnut;* chest and under-parts darker chestnut. Size the same.

Adult Female.—Probably almost similar in plumage to the
male, but no examples in which the sex is indicated have been
examined.

Range.—Western South America; Ecuador.

V. THE CHESTNUT-THROATED PARTRIDGE. ODONTOPHORUS
HYPERYTHRUS.

Odontophorus hyperythrus, Gould, P. Z. S. 1857, p. 223;
Ogilvie-Grant, Cat. B. Brit. Mus. xxii. p. 436 (1893).
Odontophorus hypospodius, Sclater and Salvin, Nomencl. Av.
Neotrop. p. 163 (1873).

Adult Male.—General colour above olive-brown, slightly
washed with rufous, and finely mottled with black; *eyebrow-
stripes, chin, throat,* and under-parts *deep rust-red*, paler towards
the middle; thighs and undertail-coverts reddish-brown mottled
with black. Total length, 11 inches; wing, 5·9; tail, 2·4;
tarsus, 2; middle toe and claw, 2·15.

Adult Female.—Differs from the male in having the breast and
rest of the under-parts *dark grey*, shading into blackish-grey on
the flanks. Measurements somewhat smaller.

Young Females.—Apparently resemble the *male parent*, and have the whole of the under-parts rust-coloured.

Range.—North-western South America; United States of Colombia.

VI. THE RUFOUS-BREASTED PARTRIDGE. ODONTOPHORUS SPECIOSUS.

Odontophorus speciosus, Tschudi, Wiegm. Arch. 1843, p. 387 ; id. Fauna Peruana, p. 281, pl. xxxiii. (1844–46); Gould, Monogr. Odontoph. pt. iii. pl. 25 (1850) ; Ogilvie-Grant, Cat. B. Brit. Mus. xxii. p. 437 (1893).

Adult Male.—Like the male of *O. hyperythrus* but distinguished by having the general colour of the top of the head and upper-parts dark reddish-brown ; the back of the neck and mantle with pale buff shaft-stripes ; *mottled black and white* eyebrow-stripes, and *black* chin and throat. Under-parts bright rust-red. Total length, 10 inches; wing, 5·7; tail, 2·2; tarsus, 1·8 ; middle toe and claw, 1·9.

Adult Female.—Differs from the *male* in having the breast and belly *dark grey.* Size rather smaller.

Range.—Western South America ; Ecuador and Peru.

VII. THE BLACK-EYEBROWED PARTRIDGE. ODONTOPHORUS MELANONOTUS.

Odontophorus melanonotus, Gould, P. Z. S. 1860, p. 382 ; Ogilvie-Grant, Cat. B. Brit. Mus. xxii. p. 438 (1893).

Adult.—General colour of the upper-parts deep brownish-black indistinctly mottled with pale rufous; eyebrow-stripes *dark brownish-black ;* chin, throat, and upper breast deep rust-red ; rest of under-parts like the back, but with more strongly marked *sandy* mottlings. Total length, 10 inches ; wing, 5·5 ; tail, 2·1 ; tarsus, 2 ; middle toe and claw, 2·15.

Nothing is known of the life history of this rare bird, and,

in the only examples I have examined, the sex had not been
ascertained, so it is uncertain if the male and female differ in
plumage.

Range.—Western South America; Ecuador.

VIII. THE BLACK-BREASTED PARTRIDGE.   ODONTOPHORUS LEUCOLÆMUS.

Odontophorus leucolæmus, Salvin, P. Z. S. 1867, p. 161;
   Ogilvie-Grant, Cat. B. Brit. Mus. xxii. p. 438 (1893).
         (*Plate XXXIIIa.*)

Adult Male and Female.—Easily distinguished from the last
species and those previously described by having the forehead,
sides of the head and throat, as well as the *chest and breast
black ;* the middle of the throat *white ;* the rest of the under
parts *rich rufous-brown* mixed with black on the belly.   In
male : Total length, 9 inches; wing, 5·3 ; tail, 2·2 ; tarsus,
1·8 ; middle toe and claw, 1·95.   *Female*, rather smaller.

Range.—Central America; Costa Rica and Veragua.

B. *Chest and breast spotted with white.*

IX. THE STARRED PARTRIDGE.   ODONTOPHORUS STELLATUS.

Ortyx (Odontophorus) stellata, Gould, P. Z. S. 1842, p. 183.
Odontophorus stellatus, Gould, Monogr. Odontoph. pt. ii. pl.
   27 (1846); Ogilvie-Grant, Cat. B. Brit. Mus. xxii. p. 439
   (1893).

Adult Male.—Above very similar to *O. guianensis*, but the
crest is longer and the hinder part bright *rufous-chestnut ;* chin
and throat *grey ;* under-parts deep *brick-red ;* the sides of the
chest and breast with *diamond-shaped white spots* sometimes
edged with black.   Total length, 10·5 inches; wing, 5·5 ; tail,
2·7 ; tarsus, 1·5 ; middle toe and claw, 1·6.

Adult Female.—Differs from the *male* in having the hinder
part of the crest deep *brownish-black.*   Rather smaller.

BLACK-BREASTED PARTRIDGE.

Range.—Central South America, extending westward to Eastern Ecuador and Eastern Peru, eastwards to Borba, Rio Madiera.

Habits.—Mr. E. Bartlett always found this species in coveys of from ten to twelve birds. In Eastern Peru he met with young birds just able to fly in the month of July.

X. THE SPOTTED PARTRIDGE. ODONTOPHORUS GUTTATUS.

Ortyx guttata, Gould, P. Z. S. 1837, p. 79.

Odontophorus guttatus, Gould, Monogr. Odontoph. pt. ii. pl. 28 (1846); Ogilvie-Grant, Cat. B. Brit. Mus. xxii. p. 439 (1893).

Odontophorus consobrinus, Ridgway, P. U. S. Nat. Mus. xvi. p. 469 (1893).

Adult Male.—Above very similar to *O. guianensis*, but the nape and mantle olive-brown instead of grey; forehead and fore-part of crest brownish-black; hinder-part bright rust-red; cheeks, chin, and throat *black with white shaft-stripes;* general colour of the under-parts *brownish-buff*, with white black-edged spots. Total length, 11·0 inches; wing, 5·9; tail, 2·8; tarsus, 1·7; middle toe and claw, 1·9.

Adult Female.—Differs from the *male* in having the whole crest brownish-black;* and well-marked *whitish-buff shaft-stripes* to the feathers of the mantle, which are scarcely visible in fully adult males. Specimens from Costa Rica and Chiriqui, where the ranges of this and the following species overlap, are somewhat intermediate, having the under-parts more or less washed with rufous.

Range.—Central America; Southern Mexico to Chiriqui.

---

* In younger examples the under-feathers of the crest are mixed with rufous.

XI. THE VERAGUA SPOTTED PARTRIDGE.   ODONTOPHORUS VERAGUENSIS.

Odontophorus veraguensis, Gould. P. Z. S. 1856, p. 107;
Ogilvie-Grant, Cat. B. Brit. Mus. xxii. p. 441 (1893).

Adult Male.—Distinguished from the *male* of *O. guttatus* by
having the whole top of the head and crest rust-red, and the
under-parts rufous or rufous-brown.   Total length, 9·5 inches;
wing, 5·7; tail, 2·5; tarsus, 1·65; middle toe and claw, 1·75.

Adult Female.—Differs only in having the top of the head and
crest browner; size slightly smaller.

Range.—Central America, extending from Panama to Costa
Rica.

XII. BALLIVIAN'S SPOTTED PARTRIDGE.   ODONTOPHORUS BALLIVIANI.

Odontophorus balliviani, Gould, P. Z. S. 1846, p. 69; id.
Monogr. Odontoph. p. iii. pl. 29 (1850); Ogilvie-Grant,
Cat. B. Brit. Mus. xxii. p. 441 (1893).

Adult.—Most like *O. veraguensis*, but the upper-parts are
more rufous and less heavily blotched with black; crest dark
chestnut; the eyebrow-stripes, chin, and a band on each side
of the head below the ear-coverts are *buff or rufous-buff;* the
throat *smoky-buff* with *pale buff shafts;* the general colour of
the under-parts *deep chestnut*, the diamond-shaped white black-
edged spots being *large* and *conspicuous*, especially on the sides.
Total length, 10·5 inches; wing, 5·8; tail, 2·7; tarsus, 1·8;
middle toe and claw, 2.

In the only specimens of this rare Partridge which I have
been able to examine, the sex had not been indicated, and
it is not known to what extent, if any, the male differs from
the female in plumage.

Range.—Western South America; Peru and Bolivia.

XIII. THE GORGETED PARTRIDGE. ODONTOPHORUS STROPHIUM.

Ortyx (Odontophorus) strophium, Gould, P. Z. S. 1843, p. 134.
Odontophorus strophium, Gould, Monogr. Odontoph. pt. i. pl.
   31 (1844); Ogilvie-Grant, Cat. B. Brit. Mus. xxii. p. 442
   (1893).

Adult.—Top of the head deep brown ; nape *deep chestnut ;*
mantle reddish-brown with *whitish shafts ;* primary quills uniform dark brown, the upper-parts otherwise very similar to
those of *O. guianensis ;* chin *white ;* throat *black ;* fore-part
of neck *white,* margined below by a *black band ;* general colour
of under-parts deep rust-red, paler towards the middle, and all
all spotted with white. Total length, 10·5 inches ; wing, 5·8 ;
tail, 2·5 ; tarsus, 2 ; middle toe and claw, 2·1.

It is not known whether the sexes differ in plumage. It is
possible that the above description may apply to the adult
female, and that the white shaft-stripes of the mantle may be
characteristic of that sex, as in *O. guttatus.*

Range.—North-western South America ; the United States of
Colombia.

XIV. THE CARACAS SPOTTED PARTRIDGE. ODONTOPHORUS
COLOMBIANUS.

Odontophorus columbianus, Gould, P. Z. S. 1850, p. 94 ; id.
   Monogr. Odontoph. pt. iii. pl. 30 (1850); Ogilvie-Grant,
   Cat. B. Brit. Mus. xxii. p. 442 (1893).

Adult.—Upper-parts very similar to those of the last species,
O. strophium, but the *throat* as well as the chin white,
barred on the sides *with black,* and surrounded by a black
band widest on the fore-neck ; the general colour of the
under-parts reddish-brown, the white spots being large and
well-marked. Total length, 11 inches ; wing, 5·7 ; tail, 2·5 ;
tarsus, 1·85 ; middle toe and claw, 2·05.

It is not known in what particulars, if any, the sexes differ
in plumage.

Range.—Northern South America ; Venezuela.

THE LONG-LEGGED COLINS. GENUS RHYNCHORTYX.

Rhynchortyx, Ogilvie-Grant, Cat. B. Brit. Mus. xxii. p. 443 (1893).

Type, *R. spodiostethus* (Salvin).

Plumage of sexes similiar (?)* No distinct crest.

First primary flight-feather *intermediate in length* between the eighth and ninth ; fourth longest.

Tail composed of *ten* feathers, less than half the length of the wing.

Tarsus *longer than* the middle toe and claw.

Bill very stout.

Only two small species are known.

I. THE LONG-LEGGED COLIN. RHYNCHORTYX SPODIOSTETHUS.

Odontophorus spodiostethus, Salvin, Ibis, 1878, p. 447.

Rhynchortyx spodiostethus, Ogilvie-Grant, Cat. B. Brit. Mus. xxii. p. 443 (1893).

(*Plate XXXIV.*)

Adult.—Crown of head rusty-brown ; mantle dark grey, edged with reddish-brown ; lower back and rump thickly mottled with grey ; shoulder-feathers brown with chestnut margins ; wing-coverts greyish-brown mixed with buff and blotched with black ; forehead, eyebrow-stripes, sides of head, and throat *bright rust-red ;* middle of chin and throat *whitish ;* neck, *chest*, and sides of breast, *dark-grey*, shading into buff on the middle of the breast and belly. Total length, 7·5 inches ; wing, 4·9 ; tail, 1·8 ; tarsus, 1·4 ; middle toe and claw, 1·15.

Range.—Central America ; Panama.

II. THE CHESTNUT-ZONED LONG-LEGGED COLIN. RHYN-CHORTYX CINCTUS.

Odontophorus cinctus, Salvin, Ibis, 1876, p. 397 ; Rowley, Orn. Misc. iii. p. 39, pl. lxxxvi. (1878).

---

* No specimens in which the sex has been ascertained have been ex-amined.

LONG-LEGGED COLIN.

Rhynchortyx cinctus, Ogilvie-Grant, Cat. B. Brit. Mus. xxii. p. 444 (1893).

Adult.—Differs from *R. spodiostethus* in having the crown, mantle, *chest*, and sides of the breast *deep rufous;* the eyebrow-stripes and sides of head *dull olive-brown ;* a white band from the base of the upper mandible to the eye, and continued behind the eye along the sides of the neck ; the chin and throat white; and the rest of the under-parts white, barred with black, except down the middle of the body.   Total length, 7·5 inches; wing, 4·6 ; tail, 1·7 ; tarsus, 1·3 ; middle toe and claw, 1·15.

Range.—Central America ; Veragua.

THE MEGAPODES, BRUSH-TURKEYS, AND MALEOS.   FAMILY MEGAPODIIDÆ.

Distinguished by having the hind toe or hallux *on the same level as the other toes*, and its basal phalanx as long as that of the third toe.

The oil-gland at the base of the tail *nude*.

Eggs deposited in the sand or in a mound raised by one or more pairs of birds ; the young hatched fully feathered, without the intervention of the parent bird, and able to fly almost from birth.

Dr. A. R. Wallace makes the following remarks on this interesting group of birds :—

"The very peculiar habits of the whole family of the *Megapodiidæ*, departing widely from those of all other birds, may also, I think, be shown to be almost the necessary results of certain peculiarities of organisation.   These peculiarities are two—the size and the number of the eggs, and the nature of the food on which these birds subsist.   Each egg being so large as to fill up the abdominal cavity and with difficulty pass the walls of the pelvis, a considerable interval must elapse before the succeeding ones can be matured.   The number of

M 2

eggs which a bird produces each season seems to be about eight, so that an interval of *three months* elapses between the laying of the first and last egg.   Now, supposing the eggs to be hatched in the oidinary way, they must be laid on the ground (for the general structure of the birds renders the construction of an arboreal nest impossible) and must be incessantly watched by the parents during that long interval, or they would be surely destroyed by the large lizards which abound in the same district.   It seems probable, however, that the eggs could not retain the vital principle for so long a time, so that the bird would have to sit on them from the commencement, and hatch them successively.   But the period of incubation is a severe tax upon all birds, even when it is comparatively short, and food easily obtained.   In this case complete incubation would be most likely impossible, because the particular species of fruits on which these birds subsist would be soon exhausted around any one locality, and both parents and offspring would perish of hunger.   If this view is correct, the *Megapodiidæ must* behave as they do.   They must quit their eggs to obtain their own subsistence ; they must bury them to preserve them from wild animals ; and each species does this is the manner which slighter modifications of structure render most convenient."

THE TRUE MEGAPODES. GENUS MEGAPODIUS.

Megapodius, Quoy and Gaimard, Voy. Uranie, p. 125 (1824).

Type, *M. freycineti,* Q. and G.

The upper tail-coverts *not reaching* to the extremity of the tail-feathers.

In some species the head and neck for the most part feathered, while in others these parts, with the exception of the crown and nape, are almost entirely naked.

Bill *slender.*

Nostrils *oval.*

First primary flight-feather *about equal to* or rather shorter than the tenth; the fifth somewhat the longest. *Secondary quills as long as the primaries.*

Tail short and rounded, composed of *twelve* feathers.

Legs (metatarsi) and feet very large and strong; fore-part of legs covered by a single row of large scales.

Claws *long and straight*, that of the middle toe being *much longer than* the upper mandible, measured from the posterior wall of the nostril to the tip.

1. THE NICOBAR MEGAPODE. MEGAPODIUS NICOBARIENSIS.

Megapodius nicobariensis, Blyth, J. As. Soc Beng. xv. pp. 52, 372 (1846); v. Pelz. Reise Novara, Vög. p. 110, pls. iv. and vi. fig. 12 [egg] (1865); Hume and Marshall, Game-Birds of India, i. p. 119, pl. (1878); iii. App. p. 428, pl. ii. (1880); Oates, ed. Hume's Nests and Eggs, Ind. B. iii. p. 449 (1890); Ogilvie-Grant, Cat. B. Brit. Mus. xxii. p. 447 (1893).

Megapodius trinkutensis, Sharpe, Ann. Mag. N. H. (4) xiii. p. 448 (1874).

Adult Male and Female.—Upper-parts dull olive-brown, the mantle *being similarly coloured* to the rest of the upper-parts, wing-coverts and outer webs of quills brighter than the back; those of the *outer primary quills being pale ochraceous ;* under-parts *pale grey*, washed with brown on the chest. Total length, 14·5 inches; wing, 9·2; tail, 2·9; tarsus, 2·7.

Younger examples have the neck feathered and the *under-parts brown* or rufous-brown.

In some examples I have examined the crown is partially or entirely naked and covered with thick black-looking skin, which has much the appearance of a scab. This peculiarity, which is not due to age, is no doubt abnormal, and possibly caused by disease.

Range.—Nicobar Islands.

Habits.—According to Mr. A. O. Hume, " The Megapode never wanders far from the sea-shore, and throughout the day it keeps in thickish jungle, a hundred yards or so above high water mark. It never, so far as I observed, emerged on to the open grass hills that form so conspicuous a feature in many of the Nicobars, but throughout the day hugged the belt of more or less dense jungle that in most places, along the whole coast line, supervenes abruptly on the white coral beach. At dusk, during moonlight nights, and in the early dawn, glimpses may be caught of them running about on the shore or even at the very water's edge, but during daylight they skulk in the jungle.

"They are to be met with in pairs, coveys, and flocks of from thirty to fifty. They run with great rapidity and rise unwillingly, running and flying just like jungle hens. They often call to each other, and when a party has been surprised and dispersed, they keep on talking to each other incessantly, half-a-dozen cackling at the same time. The note is not unlike the chuckling of a hen that has recently laid an egg, and is anxious to publish the stupendous fact in nature's pages ; it may be syllabled in a variety of ways, but several of us agreed that on the whole kŭk-ă-kŭk-kŭk ! most nearly represented their chuckling, cackling call.

" The stomachs of all we examined contained tiny land-shells (sometimes with the animals not yet dead), larvæ of insects, dissolved matter, apparently vegetable, and minute fragments and particles of quartz or other hard rocks.

" As game they are unsurpassed. The flesh, very white, very sweet and juicy, loaded with fat, is delicious, a sort of *juste milieu* between that of a fat Norfolk Turkey and a fat Norfolk Pheasant.

" The eggs, too, are quite equal, if not superior, to those of the Pea-Fowl, and, to my mind, higher commendation cannot be given.

" But it is in regard to their nidification that these birds possess the highest interest. Moderate-sized birds as they are, they gradually manage to accumulate tumuli that would not have done discredit to the final resting-place of some ancient British hero, and in these they bury their eggs and leave them to be hatched by the heat evolved, as I believe, by fermentation in the interior of these mounds."

The late Mr. W. R. Davison, one of the finest field naturalists that ever lived, says :—" I have seen a great many mounds of this bird. Usually they are placed close to the shore, but on Bompoka and on Katchall I saw two mounds some distance inland in the forest. They were composed of dried leaves, sticks, &c., mixed with earth, and were very small compared with others near the sea-coast, not being above three feet high, and about twelve or fourteen feet in circumference; those built near the coast are composed chiefly of sand mixed with rubbish, and vary very much in size, but average about five feet high and thirty feet in circumference; but I met with one exceptionally large one on the Island of Trinkut, which must have been at least eight feet high and quite sixty feet in circumference. It was apparently a very old one, for from near its centre grew a tree about six inches in diameter, whose roots penetrated the mound in all directions to within a foot of its summit, some of them being nearly as thick as a man's wrist. I had this mound dug away almost to the level of the surrounding land, but only got three eggs from it, one quite fresh, and two in which the chicks were somewhat developed.

" Off this mound I shot a Megapode, which had evidently only just laid an egg. I dissected it, and from a careful examination it would seem that the eggs are laid at long intervals apart, for the largest egg in the ovary was only about the size of a large pea, and the next in size about as big as a small pea. These mounds are also used by reptiles, for out of one I dug, besides the Megapode's eggs, about a dozen eggs of some large lizard.

" I made careful enquiries among the natives about these

birds, and from them I learnt that they usually get four or five
eggs from a mound, but sometimes they get as many as ten ;
they all assert that only one pair of birds are concerned in the
making of a mound, and that they only work at night.   When
newly made, the mounds (so I was informed) are small, but
are gradually enlarged by the birds. . . . .

The eggs are usually buried from three-and-a-half to four
feet deep, and how the young manage to extricate themselves
from the superincumbent mass of soil and rubbish seems a
mystery.   I could not obtain any information from natives on
this point, but most probably they are assisted by their parents,
if not entirely freed by them, for these latter, so the natives
affirm, are always to be found in the vicinity of the mounds
where their eggs are deposited.

" The surface soil of the mounds only is dry; at about a
foot from the surface the sand feels slightly damp and cold, but
as the depth increases the sand gets damper, but at the same
time increases in warmth."

Eggs.—Very large elongate ovals, uniform in colour and of
three distinct types, dull clayey-pink, earthy-yellow, and earthy-
brown of several shades.   Average measurements, 3·25 by
2·07 inches.

II. THE TENIMBER MEGAPODE.   MEGAPODIUS TENIMBERENSIS.

Megapodius tenimberensis, Sclater, P. Z. S. 1883, p. 57; Ogilvie-
Grant, Cat. B. Brit. Mus. xxii. p. 448 (1893).

Adult Male.—Closely resembles *M. nicobariensis,* but may be
recognised by having the outer webs of the outer primary
quills *pale brown ;* the upper part of the mantle may be *very
slightly* washed with grey and the sides of the head, chin, and
throat are more thickly covered with small feathers.   Total
length, 14 inches ; wing, 9·6 ; tail, 3·5 ; tarsus, 2·9.

The adult female is no doubt similar to the male, but the

two typical examples in the British Museum collection are both males.

Range.—Tenimber Islands, Moluccas.

III. CUMING'S MEGAPODE. MEGAPODIUS CUMINGI.

Megapodius cumingii, Dillwyn, P. Z. S. 1851, p. 118, pl. 39;
Motley and Dillwyn, Contr. N. H. Labuan, p. 32, pl. 7
(1855); Ogilvie-Grant, Cat. B. Brit. Mus. xxii. p. 449
(1893).
Megapodius gilbertii, Gray, P. Z. S. 1861, p. 289.
Megapodius lowi, Sharpe, P. Z. S. 1875, p. 111.
Megapodius pusillus, Tweedd. P. Z. S. 1877, p. 765, pl. lxxviii.
(Juv.).
Megapodius dillwyni, Tweedd. P. Z. S. 1877, p. 766.

Adult Male and Female.—Whole plumage darker than in *M. nicobariensis*, especially the under-parts, which are *dark* grey. The upper-parts vary considerably in different individuals, some being olive-brown, others more or less washed with rufous. Total length, 15 inches; wing, 9; tail, 3; tarsus, 2·7.

Range.—Philippine Islands, extending south to Palawan and the small islands off the north coast of Borneo; also met with in the Sula Islands, Celebes, and Tojian Islands.

Habits.—Messrs. Motley and Dillwyn give the following notes on Cuming's Megapode:—"In Labuan they are not uncommon, and are said to be principally confined to small islands, to such more especially as have sandy beaches; they are very rarely to be seen, being extremely shy and frequenting dense and flat parts of the jungle, where the ratans grow, and where the luxuriance of the vegetation renders concealment easy. The Malays snare them by forming long thick fences in unfrequented parts of the jungle, in which at certain intervals they leave openings where they place traps; the birds run through the jungle in search of food, and coming to this fence, run

along it till they find one of the openings, through which they
push their way, and are caught in the trap.   In walking they
lift up their feet very high, and set up their backs something
like Guinea-fowls ; they frequently make a loud noise like the
screech of a chicken when caught ; they are very pugnacious,
and fight with great fury by jumping upon one another's backs
and scratching with their long strong claws.   Their food
principally consists of seeds and insects." A very large and
perfect mound about twenty feet in diameter was visited
by Mr. Motley, and was composed of sand, earth, and sticks,
and situated just within the jungle above high-water mark.
The boatmen managed to find about a dozen eggs buried at a
depth of from one to three feet and placed in an upright position, the ground about them being astonishingly hard.   The
eggs thus obtained were placed in a box of sand, and it was
afterwards discovered that they had all hatched, but from
neglecting to place them in a proper (*i.e.*, probably upright)
position, the chicks had been unable to get up through the
sand and had all perished.   On another occasion one of a number of eggs brought in by natives hatched out at the end of three
weeks.   A Malay who saw the young bird emerge said that it
just shook off the sand and ran away so fast that it was only
caught with difficulty ; it then appeared to be nearly half-grown,
and from the first fed itself without hesitation, scratching and
turning up the sand like an old bird.

Eggs.—Like those of *M. nicobariensis* ; long, perfect ovals ;
pinkish stone colour. Average measurements, 3·2 by 2
inches.

IV. THE SANGHIR MEGAPODE.   MEGAPODIUS SANGHIRENSIS.

Megapodius sanghirensis, Schlegel, Notes Leyd Mus. ii. p. 91
    (1880); Ogilvie-Grant, Cat. B. Brit. Mus. xxii. p. 450
    (1893).

Adult.—Upper-parts *dark chestnut-brown* without any olive

shade; under-parts *dark brown*. Total length, 14 inches; wing, 9; tail, 3; tarsus 2·5.

Range.—Sanghir Islands.

V. BERNSTEIN'S MEGAPODE. MEGAPODIUS BERNSTEINI.

Megapodius bernsteinii, Schlegel, Ned. Tijdschr. Dierk. iii. p. 261 (1866); Ogilvie-Grant, Cat. B. Brit. Mus. xxii. p. 450 (1893).

Adult.—Upper-parts much like those of *M. cumingi*, but the outer webs of the flight-feathers more strongly washed with rufous and the lower back dark brownish-chestnut; chest olive-brown, shading into *rufous-brown* on the under-parts, which are not nearly so dark as those of *M. sanghirensis*. Total length, 12 inches; wing, 7·5; tail, 2·3; tarsus, 2·5.

Range.—Sula Islands, Celebean Archipelago.

VI. FORSTEN'S MEGAPODE. MEGAPODIUS FORSTENI.

Megapodius forstenii, Temm.; Gray, Gen. B. iii. p. 491, pl. 124 (1847); Ogilvie-Grant, Cat. B. Brit. Mus. xxii. p. 451 (1893).
Megapodius affinis, Meyer, SB. Ak. Wien. lxix. p. 215 (1874).
Megapodius decollatus, Oustal. Bull. Assoc. Sci. Fr. xxi. p. 248 (1878).

Adult.—Mantle *dark grey*, *contrasting* with the olive-brown of the rest of the upper-parts; feathers of the forehead *extending to the base of the bill*; a short but distinct crest; belly brownish-grey; legs *dark* (olive-brown). Total length, 14·5 inches; wing, 8·6; tail, 2·8; tarsus, 2·7.

Range.—Bourou, Amboina, Ceram, and Goram, extending North to Western New Guinea, Jobi Island, and d'Urville Island off the north coast of New Guinea.

Habits.—Dr. A. R. Wallace notes that "this bird deposits its eggs in a heap of rubbish collected in low places near the sea.

It is semi-nocturnal in its habits, making a loud wailing cry, which is often heard at night and about daybreak."

VII. THE ASTROLABE BAY MEGAPODE.   MEGAPODIUS BRUNNEIVENTRIS.

Megapodius brunneiventris, Meyer, Abh. Zool. Mus. Dresd. 1890–91, No. 4, p. 15 (1892) ; Ogilvie-Grant, Cat. B. Brit. Mus. xxii. p. 452 (1893).

Adult.—Said to resemble *M. forsteni*, but with the belly brown. I have had no opportunity of examining the typical example of this species, but it seems probable that it may prove to be merely an immature example of Forsten's Megapode.

Range.—Astrolabe Bay, North-east New Guinea.

VIII. BRENCHLEY'S MEGAPODE.   MEGAPODIUS EREMITA.

Megapodius eremita, Hartlaub, P. Z. S. 1867, p. 830 ; Sclater, Voy. Challenger, p. 32, pl. xi. (1880) ; Studer, in Forschungsreise, S. M. S. "Gazelle," iii. p. 253, pl. xii. (1889) ; Ogilvie-Grant, Cat. B. Brit. Mus. xxii. p. 452 (1893).

Megapodius brenchleyi, Gray, Ann. Mag. N. H. (4) v. p. 328 (1870); id. Cruise "Curaçoa," p. 392, pl. 20 (1873); Sharpe, in Gould's B. New Guin. pt. xxii. pl. xi. (1886).

Megapodius hueskeri, Cab. and Reichenow, J. f. O. 1876, p. 326.

Megapodius rubrifrons, Sclater, P. Z. S. 1877, p. 556.

Adult Male and Female.—Forehead *naked*, with the exception of a few minute plumes ; mantle *dark grey*, contrasting with the olive-brown head and upper-parts ; lower back washed with dark chestnut ; under-parts dark grey, browner on the belly and thighs. Legs *black, or dark olive*. Total length, 15 inches ; wing, 8·8–9·2 ; tail, 2·5–3 ; tarsus, 2·6–2·8.

Range.—Admiralty Islands, New Hanover, New Ireland, Duke of York Island, New Britain and the Solomon Islands.

Habits.—Writing of this Megapode as observed in New Britain, where it is especially numerous, Mr. E. L. Layard says:—"This bird is a perfect nuisance in Blanche Bay, the whole place, both on the grassy flats and bush-covered hill-sides, being so undermined with their nesting-holes that we were continually stumbling into them, notwithstanding all our care in walking. Like domestic Fowls, they lay indiscriminately in each other's nests. Some of these are regular excavations, six or seven feet deep. Going shooting one day, I saw two flat white things moving in the mouth of a small cavern by the side of the road. Upon closer inspection they proved to be the upturned soles of a native's feet, their owner being head downwards, nearly six feet underground. He presently emerged with five eggs, which I purchased on the spot for a penny stick of tobacco. The consumption of eggs by the 'Renard's' thirty men was something enormous, the price alongside being six eggs for one stick of tobacco. The birds were very numerous, and when flushed took to the trees. . . . The great holes in the plain are easily accounted for. A Megapode scratches a hole and buries her egg; a native comes along, rakes out the egg with his hands, but does not fill the hole up again. Another bird lays at the bottom of the excavation, and a native digs it out again, until at length a perfect tunnel is formed in the soft volcanic earth."

According to Dr. O. Finsch:—"They seem to lay all the year round, except in the rainy months, when eggs are very rare, and for a short time not to be had at all. A year and a half ago (in 1881), forty eggs could be bought from the natives for one stick of tobacco; now one gets only two.

"Although *Megapodius eremita* is very common, one very seldom gets a sight of it. It runs very quickly through the jungle, or is seen only on the wing for a moment among the foliage of the trees. The young ones, when hatched, are already able to fly. It is singular that among the numerous specimens (about forty) I got, there was not a single male bird; they were all females."

Eggs.—Like those of *M. nicobariensis*, pale cinnamon colour. Average measurements, 3 by 1·9 inches.

IX. MACGILLIVRAY'S MEGAPODE.   MEGAPODIUS MACGILLI-
VRAYI.

Megapodius macgillivrayi, Gray, P. Z. S. 1861, p. 289; 1864, p. 43; Sclater, P. Z. S. 1876, p. 460, pl. xliii.; Ogilvie-Grant, Cat. B. Brit. Mus. xxii. p. 453 (1893).

Adult Male and Female.—Much like *M. eremita* in general appearance, but easily distinguished by the colour of the legs, which are *orange-red;* the feathers on the crown and back of the head are rather long, and form *a short thick crest;* flanks *blackish-grey.* Total length, 13·5 inches; wing, 8·5–8·9; tail, 3–3·2; tarsus, 2·3–2·7.

Range.—Louisiade Archipelago, extending to the shores of Huon Gulf and the Astrolabe Mountains.

Habits.—Mr. J. Macgillivray, the original discoverer of this Megapode, writes as follows:—"In habits this bird resembles the Australian species, especially in constructing enormous mounds for the reception of its eggs. Those which I saw averaged five feet in height and fifteen in diameter, and were composed of the sandy soil of the neighbourhood, mixed up with rotten sticks and leaves, but without any shells or coral. Some were placed on the outer margin of the thickets close to the beach, and others were scattered about more inland. As several of these mounds showed indications of having lately been opened by the birds, I entertained hopes of being able to procure an egg, but after digging several pits three feet in depth, with no more efficient implements than my hands, I had to give up the work from sheer exhaustion. This bird is apparently very pugnacious at times, as I frequently saw them chasing each other along the ground, running with great swiftness, and uttering their cry more loudly than usual, stopping short suddenly, and again starting off in pursuit. The cry consists of one or two shrill notes uttered at intervals, and ending in a hurried tremulous cry, repeated five or six times. The noise made by this *Megapodius* while scratching among the dead leaves for food may sometimes be imitated with such success as to bring the bird running up within gunshot. When suddenly forced to rise from the ground, it flies up into a tree, and remains there motionless, but exceedingly vigilant, ready to start on the approach of anyone, but on other occasions it trusts to its legs to escape. Its food is entirely procured on the ground, and consists of insects and their larvæ (especially the pupæ of ants), small snails, and various fallen seeds and fruits."

X. DUPERREY'S MEGAPODE.   MEGAPODIUS DUPERREYI.

Megapodius duperreyii, Less. and Garn. Bull. Sci. Nat. viii. p. 113 (1826); Less. Voy. "Coquille," i. pt. ii. p. 700, pl. 36 (1828); Ogilvie-Grant, Cat. B. Brit. Mus. xxii. p. 454 (1893).

Megapodius rubripes, Temm. Pl. Col. v. pl. 46 [No. 411] (1826).

Megapodius reinwardtii, Wagler, Syst. Av. Addit. Megapodius, p. 378, sp. 4 (1827); id. Isis, 1829, p. 736.

Megapodius tumulus, Gould, P.Z.S. 1842, p. 20; id. B. Austr. v. pt. vi. pl. 79 (1842); Gray, P. Z. S. 1861, p. 290, pl. xxxiv.

Megapodius gouldi and *M. amboinensis*, Gray, P. Z. S. 1861, pp. 290, 293.

Megapodius assimilis, Masters, P. Linn. Soc. N. S. W. i. p. 59 (1887).

Adult Male and Female.—A well-developed brown crest; mantle *grey ;* back and wings olive-brown; lower back and rump dark chestnut; under-parts smoky-grey; sides and *flanks* mostly

dark chestnut. Legs orange-red. Total length, 14–16·8 inches ;
wing, 8·8–10·2 ; tail, 3·6–4·1 ; tarsus, 2·55–3·0.

Range.—Extending in the west to the Kangeang Archipelago
and thence eastwards through Lombock, Flores, Sumba,
Banda, Kei, Aru, Salawatti, Western and Southern New
Guinea, and the islands of Torres Straits to North-east
Australia.

As might be expected with a species occupying so wide a
range, considerable differences are to be found on comparing
examples from Flores and Lombock (*M. gouldi*, Gray) with
those from Australia, the former being smaller and much
lighter in colour than the latter. But these differences are so
entirely bridged over by specimens from the intermediate
islands, that it is impossible to regard them as being specifi-
cally distinct, the two extremes passing imperceptibly into
one another.

Habits.—The following interesting notes are extracted from
Gilbert's account of the habits of this species as observed by
him in Northern Australia. He says :—" I landed beside a
thicket, and had not proceeded far from the shore ere I came
to a mound of sand and shells, with a slight mixture of black
soil, the base resting on a sandy beach, only a few feet above
high water mark ; it was enveloped in the large yellow-blos-
somed *Hibiscus*, was of a conical form, twenty feet in circumference at the base, and about five feet in height. On point-
ing it out to a native and asking him what it was, he replied,
" Oooregoorgā Rambal," Megapode's house or nest. I then
scrambled up to the sides of it, and to my extreme delight
found a young bird in a hole about two feet deep ; it was lying
on a few dry, withered leaves, and appeared to be only a few
days old. . . .

" As it fed rather freely on bruised Indian corn, I was in
full hopes of rearing it ; but it proved of so wild and in-
tractable a disposition that it would not reconcile itself to such

close confinement, and effected its escape on the third day. During the period it remained in captivity it was incessantly occupied in scratching up the sand into heaps; and the rapidity with which it threw the sand from one end of the box to the other was quite surprising for so young and small a bird, its size not being larger than that of a small Quail. At night it was so restless that I was constantly kept awake by the noise it made in its endeavours to escape. In scratching up the sand it only used one foot, and having grasped a handful, as it were, the sand was thrown behind it, with but little apparent exertion, and without shifting its standing position on the other leg. . . .

"I continued to receive the eggs without having an opportunity of seeing them taken from the mound until the 6th of February, when on again visiting Knocker's Bay I had the gratification of seeing two taken from a depth of six feet in one of the largest mounds I had then seen. In this instance the holes ran down in an oblique direction from the centre towards the outer slope of the hillock, so that, although the eggs were six feet deep from the summit, they were only two feet deep from the side. The birds are said to lay but a single egg in each hole, and after the egg is deposited the earth is immediately thrown in lightly until the hole is filled up; the upper part of the mound is then smoothed and rounded over. It is easily known when a Megapode has been recently excavating, from the distinct impressions of its feet on the top and sides of the mound, and the earth being so lightly thrown over, that with a slender stick the direction of the hole is readily detected, the ease or difficulty of thrusting the stick down indicating the length of time that may have elapsed since the bird's operations. . . .

"I revisited Knocker's Bay on the 10th of February, and having with some difficulty penetrated into a dense thicket of cane-like, creeping plants I suddenly found myself beside a mound of gigantic proportions. It was fifteen feet in height and

sixty in circumference at the base, the upper part being about
a third less, and was entirely composed of the richest description of light vegetable mould; on the top were very recent
marks of the bird's feet. The native and myself immediately
set to work, and after an hour's extreme labour, rendered the
more fatiguing from the excessive heat, and the tormenting
attacks of myriads of mosquitoes and sand-flies, I succeeded in
obtaining an egg from a depth of about five feet; it was in a
perpendicular position, with the earth surrounding and very
likely touching it on all sides, and without any other material
to impart warmth, which in fact did not appear necessary,
the mound being quite warm to the hands. The holes in this
mound commenced at the outer edge of the summit, and ran
obliquely towards the centre; their direction, therefore, is not
uniform. Like the majority of the mounds I have seen, this
was so enveloped in thickly foliaged trees as to preclude the
possibility of the sun's rays reaching any part of it. . .
The mounds are doubtless the work of many years, and of
many birds in succession; some of them are evidently very
ancient, trees being often seen growing from their sides; in one
instance I found a tree growing from the middle of a mound
which was a foot in diameter. . . . The natives say that
only a single pair of birds are ever found at one mound at a
time, and such, judging from my own observation, I believe
to be the case; they also affirm that the eggs are deposited at
night, at intervals of several days, and this I also believe to be
correct, as four eggs taken on the same day, and from the same
mound, contained young in different stages of development;
and the fact that they are always placed perpendicularly is
established by the concurring testimony of all the different
tribes of natives I have questioned on the subject. . . .
It is at all times a very difficult bird to procure; for although
the rustling noise produced by its stiff pinions when flying may
be frequently heard, the bird itself is seldom to be seen. Its
flight is heavy and unsustained in the extreme; when first dis-

turbed it invariably flies to a tree, and on alighting stretches out
its head and neck in a straight line with its body, remaining in this position as stationary and motionless as the branch
upon which it is perched; if, however, it becomes fairly
alarmed, it takes a horizontal but laborious flight for about
a hundred yards, with its legs hanging down as if broken."

Eggs.—Very large ; in shape, long perfect ovals; pale coffee-
brown when newly laid, but after remaining in the mound a
few days they become darker. Average measurement, 3·55 by
2·1.

XI. FREYCINET'S MEGAPODE. MEGAPODIUS FREYCINETI.

Megapodius freycinet, Quoy and Gaim. Voy. "Uranie," p. 125,
pl. 32 (1824).
Megapodius freycineti, Temm. Pl. Col. v. pl. 45 [No. 220]
(1825?); Ogilvie-Grant, Cat. B. Brit. Mus. xxii. p. 457
(1893).
Alecthelia urvilii, Less. and Garn. Bull. Sci. Nat. viii. p. 115
(1826).
Megapodius quoyi, Gray, P. Z. S. 1861, p. 289, pl. xxxii.

Adult Male and Female.—*General plumage black* washed with
grey, except the wings, lower back, and belly, which are
tinged with brown ; feathers on the head extending to the base
of the bill and forming a short full crest. Legs *black*. Total
length, 15 inches ; wing, 8·7–9 ; tail, 3–3·3 ; tarsus, 2·7–3.

Range.—The Molucca Islands and Western New Guinea.

XII. THE GEELVINK BAY MEGAPODE. MEGAPODIUS GEELVINKIANUS.

Megapodius geelvinkianus, Meyer, SB. Ak. Wien, lxix, p. 88
(1874); Ogilvie-Grant, Cat. B. Brit. Mus. xxii. p. 459
(1893).
Megapodius affinis, Salvad. Ann. Mus. Civ. Genov. viii. p. 406
(1876).

N 2

Adult.—Allied to *M. freycineti*, but said to be distinguished by having the legs cherry-red. Total length, 14 inches; wing, 8·5–9·5; tail, 2·7–3; tarsus, 2·5–2·8. I have not seen an example of this species.

Range.—Mafoor, Misori, and Jobi Islands in Geelvink Bay, and Dorey, New Guinea.

XIII. LAYARD'S MEGAPODE.   MEGAPODIUS LAYARDI.

Megapodius brazieri, Sclater, P. Z. S. 1869, p. 528 [founded on an egg from Banks I.].
Megapodius layardi, Tristram. Ibis, 1879, p. 194; Ogilvie-Grant, Cat. B. Brit. Mus. xxii. p. 459 (1893).

(*Plate XXXV.*)

Adult Male and Female.—Head and neck with the exception of the back of the head and nape *almost naked ;* general plumage *black*, washed with grey, except the wings and rump, which have an olive tinge ; belly paler than the breast. Legs *yellow*. Total length, 17 inches; wing, 9·5 ; tail, 3·5 ; tarsus, 2·9.

Range.—New Hebrides.

Mr. Layard writes:—"The native name is 'Malou.' This bird is getting very scarce in consequence of the rapid increase of pigs and tame cats that have taken to the bush. It is a very shy and wary bird, and is found only on the sides of deep, densely wooded ravines, where it scratches about among the rocks for the worms, small helices, and little hard seeds that form its food. The natives among the New Hebrides group tell me that in their islands the 'Malou' deposits its eggs in a hole scratched under a rotten fallen log in the forest, and then covers them up with leaves. This account was confirmed by an intelligent missionary on the island of Sandwich, or Vate. . . . While staying on Vate I offered a large reward in beads, tobacco, and tomahawks to any native who would conduct me to a nest, so that I could get the eggs out with my own hands. Just two days after I had left in the *Dayspring* for

LAYARD'S MEGAPODE.

the other islands, a man brought three eggs, fresh laid. He was told to come back again as soon as the vessel returned; but he did not, and I never saw a nest. I travelled to a place on the eastern side of Vate, where I was told there were still a few of these birds remaining. . . . I stole along carefully, just stepping from one rock to another, and every few yards stopping behind a tree to listen and reconnoitre. . . . Twice, I was certain, I heard scratching among the dead leaves, but could see no birds. I could have had several shots at fowls run wild, but I was after nobler game. At length, as the bats were already flitting around my head, I thought it time to retrace my footsteps. I had not gone far, when, with a hoarse croak, a dark object bounded over the bottom of the watercourse I was walking in. In the gathering darkness I could only see a black mass, like a stone, among the saplings. However, as I knew I could not get any nearer, I tried the chokebore at it. The smoke hung round so that I could see nothing, and I heard no fluttering among the leaves; but when I went up to the place there lay my first 'Malou,' shot through the head and heart. A little further on I heard the scratching (sure sign!); but while cautiously peeping round a big tree, an envious rotten branch caught against my breast, and broke with a loud snap; and I just got a glimpse of the 'Malou' running like a racehorse over a slight elevation close by. Next day I was in a ravine so precipitous that I had to get into the summit of a big tree and climb down that way. I had not gone far when I heard something that sounded remarkably like a 'Malou.' As before, I was in the bed of the watercourse. I looked all about the almost perpendicular sides. There was nothing to be seen, but the noise still continued; and at last, right in front of me, on a little pebbly bank under a huge rock, I caught sight of two splendid 'Malous,' slowly retreating, and looking full at me. They were evidently old birds, in full breeding plumage, their bare red heads and necks shining grandly in a gleam of sunshine; and they carried their absurd little tails stuck straight

down between their legs. I was delighted at getting such a
good look at so rare a bird, and tried to get both at one shot ;
but I have been 'sold' too often by being too greedy, so
knocked over the furthest one with a half-charge. The other
one apparently vanished into thin air, as I saw nothing more
of him. About a mile higher up the ravine I was startled by
the hoarse cry of alarm, which appears to be the only sound
these birds emit; and I could just see the bird's red head as
he stared at me from behind a clump of rocks. I soon had
the pleasure of handling him."

Eggs.—Similar to the eggs of *M. nicobariensis*, but, as a rule,
rather longer and more pointed ovals. Average measurements,
3·25 by 1·9 inches.

XIV. THE PELEW MEGAPODE. MEGAPODIUS SENEX.

Megapodius senex, Hartlaub, P. Z. S. 1867, p. 830; Finsch,
    J. Mus. Godeffr. iv. pt. viii. p. 29, pl. v. figs. 2–3 (1875).
Megapodius laperousii, Ogilvie-Grant, Cat. B. Brit. Mus. xxii.
    p. 460 (1893) [part. ; Pelew Is.]

Adult.—General colour of plumage greyish-black, washed
with dark olive-brown on the wings and deep reddish-brown
on the rump ; top of the head *French grey ;* forehead, sides
of the head, and throat similar, but thinly covered with feathers.
Bill and legs *yellow*. Total length, 9·5 inches ; wing, 7 ; tail,
2·3 ; tarsus, 2.

Range.—Pelew Islands.

Since writing on these birds in the "Catalogue of Birds,"
Mr. Hartert has shown me a number of *M. laperousii* from the
Marianne Islands, in the Hon. Walter Rothschild's collection.
As he points out, Dr. Oustalet was evidently wrong in uniting
this species with *M. senex* from the Pelew Islands, as the dif-
ferences between them, though slight, appear to be constant.

Eggs.—Like those of the other species. Measurements, 2·9
by 1·8.

XV. THE MARIANNE MEGAPODE. MEGAPODIUS LAPEROUSII.

Megapodius la Pérouse, Quoy and Gaimard, Voy. "Uranie,"
p. 127, pl. 33 (1824).

Megapodius laperousii, Temm. Pl. Col. v. livr. 69, p. 3 (1826);
Ogilvie-Grant, Cat. B. Brit. Mus. xxii. p. 460 (1893) [part. ;
Marianne Is.].

Megapodius perousii, Wiglesw. Abh. Mus. Dresd. 1890-91, No.
6, p. 58 (1892).

Adult Male and Female.—May be distinguished from *M. senex*
by having the crown *much darker grey*, while the posterior band
of feathers above the nape are *almost white*, and form a marked
contrast with the crown.

Range.—Marianne Islands.

XVI. PRITCHARD'S MEGAPODE. MEGAPODIUS PRITCHARDI.

Megapodius stairi and *M. burnabyi*, Gray, P. Z. S. 1861, p. 290
[founded on eggs].

Megapodius pritchardi, Gray, Ann. Mag. N. H. (3) xiv. p. 378
(1864); id. P. Z. S. 1864, p. 41, pl. vi.; Ogilvie-Grant,
Cat. B. Brit. Mus. xxii. p. 461 (1893).

Megapodius huttoni, Buller, Trans. N. Zeal. Inst. iii. p. 114
(1870).

Adult.—Distinguished from all the species previously mentioned by having the *basal part of the primary and outer
secondary quills white ;* the longer upper tail-coverts and tail
mixed with the same colour.* General colour of the rest of
the plumage dull lead-grey, shading into yellowish-grey on the
belly and under tail-coverts, only the wings and back being
rufous-brown ; legs pale red. Total length, 10·5 inches ; wing,
7·4 ; tail, 2·15 ; tarsus, 2·1.

Range.—Ninafou or Hope Island.

Habits.—Captain McLeod, who visited the island of Ninafou,

* The amount of white appears to vary in individual examples.

says that the bird lives in the scrubs in the centre of the island,
about a large lagoon of brackish water, which has the appearance of being an extinct crater; the birds lay their eggs on
one side only of this lagoon, where the soil is composed of
sulphur-looking sand ; the eggs are deposited from one to two
feet beneath the surface. The locality frequented by these
birds is, on this island, under the protection of the king or
chief; and by his permission only can the birds or eggs be
procured. The number of eggs deposited in the mounds
varies, as the eggs are laid by different birds in succession ;
but as many as forty eggs are said to have been procured from
one mound.

Mr. F. Hübner also makes the following observations : —
" The breeding time of this species is not so confined to
certain months as has been noticed by Dr. Wallace in respect
of certain Malayan species. He gives as the season of incubation August and September ; but of this bird I got fresh
eggs in October and November also, and, according to Captain
Nagel and the natives, eggs are to be found likewise in other
months. Immediately after leaving the eggs, the young birds
are not only able to run, but also to fly. The old birds are
excellent runners, but their flight is somewhat heavy, as in the
common fowl; when alarmed they perch on trees. The
stomachs of those I shot were mostly filled with land-shells,
small crabs, and scolopendras, but in a few cases I found
seeds. . . . The male may be distinguished at once from
the female by its orange feet, which in the latter are yellow."

Eggs.—Similar to those of the other species already described,
but more than usually pointed ovals, and rather smaller
Average measurements, 2·9 by 1·7 inches.

THE PAINTED MEGAPODES. GENUS EULIPOA.
Eulipoa, Ogilvie-Grant, Cat. B. Brit. Mus. xxii. p. 462 (1893).
Type, *E. wallacii* (Gray).

Like *Megapodius*, but the *secondary quills are much shorter
than the primaries.*

WALLACE'S PAINTED MEGAPODE.

Head and neck *feathered*, with the exception of a small space round the eye.

Upper tail-coverts much shorter than the tail-feathers.

First primary flight-feather intermediate in length between the seventh and eighth, and *much longer than* the tenth.

Only one species is known.

I. WALLACE'S PAINTED MEGAPODE.   EULIPOA WALLACII.

Megapodius wallacii, Gray, P. Z. S. 1860, p. 362, pl. clxxi.

Eulipoa wallacii, Ogilvie-Grant, Cat. B. Brit. Mus. xxii. p. 462 (1892).

(*Plate XXXVa.*)

Adult.—General colour of the upper-parts olive; short crest rufous-olive; inner median wing-coverts, shoulder-feathers, and middle of the back slate-grey, tinged with olive, and widely barred with bright chestnut, the former tipped with pale whitish-grey; greater secondary-coverts olive, and similarly banded with chestnut; lower back, rump, and under-parts dark grey, except the middle of the belly, which is white. Total length, 14 inches; wings, 7·6–8; tail, 2·6–3·2; tarsus, 2·2.

Range.—Islands of Gilolo, Batchian, Ternate, Bourou, Ceram, and Amboina.

Habits.—"This species," writes Dr. A. R. Wallace, " differs somewhat in its habits from the other members of the Family found in the Malay Islands. It resides generally in the hilly districts of the interior, like the Maleo (*Megacephalon maleo*), and, like that species, comes down to the beach to deposit its eggs; but instead of scratching a hole for them and covering it up again, the bird burrows into the sand to a depth of three or four feet obliquely downwards, and deposits its egg at the bottom. It then loosely covers up the mouth of the hole, and is said by the natives to obliterate and disguise, by innumerable tracks and scratches, its own footmarks leading to the hole. Its offspring is then left to make its way into the world as best it can. This bird lays its eggs only at night, and the only specimen I obtained here (the Island of Bourou) was caught on the beach, at the mouth of its burrow, early one morning. Its wing was broken and wounded at the outer joint, as if it had been attacked by some small animal when in its burrow, probably a rat."

The same writer also remarks: " All these birds seem to be semi-nocturnal, for their loud, wailing cries may be constantly heard late into the night, and long before daybreak in the morning."

Eggs.—" Rusty-red ; very large for the size of the birds, being generally 3 or 3¼ inches long by 2 or 2¼ wide." (*A. R. Wallace.*)

THE OCELLATED MEGAPODES. GENUS LIPOA.

Leipoa, Gould, P. Z. S. 1840, p. 126.

Type, *L. ocellata*, Gould.

Easily distinguished from *Megapodius* and the other allied forms by having the longer upper tail-coverts *reaching to the end of the tail*.

Feathers of the top of the head forming a short thick crest.

Nostrils elongate ovals.

First primary flight-feather intermediate in length between the ninth and tenth ; fifth slightly the longest.

Tail long and rounded, composed of *sixteen* feathers.

Legs (metatarsi) and toes rather short ; the former with a double row of large hexagonal plates down the front.

Only one species is known.

I. THE OCELLATED MEGAPODE. LIPOA OCELLATA

Leipoa ocellata, Gould, P. Z. S. 1840, p. 126 ; id. B. Austr. v. pt. 1, pl. 78 (1840) ; North, Nests and Eggs Austr. B. p. 281 (1889).

Lipoa ocellata, Ogilvie-Grant, Cat. B. Brit. Mus. xxii. p. 463
(1893).

Adult Male and Female.—Top of the head dark brown, the
feathers forming a thick, pointed crest; mantle mostly grey;
back, shoulder-feathers, and wing-coverts grey, widely banded
with brownish-black and white; inner secondary quills very
similar; forehead and eyebrow-stripes greyish; cheeks and
throat rust colour; upper breast grey, with a band of black and
white feathers down the middle; rest of under-parts whitish,
barred with black on the sides. Total length, 24 inches; wing,
12·5; tail, 9; tarsus, 3·1.

Range.—South and West Australia.

Habits.—Mr. Gilbert gives the following account of this handsome bird's habits :—" This morning I had the good fortune
to penetrate into the dense thicket I had been so long anxious
to visit in search of the Leipoa's eggs, and had not proceeded
far before the native who was with me told me to keep a good
look-out, as we were among the *Ngou-oo's* hillocks; and in half
an hour after we found one, around which the brush was so
thick that we were almost running over it before seeing it.
. . . He began scraping off the earth very carefully from
the centre, throwing it over the side, so that the mound very
soon presented the appearance of a huge basin; about two feet
in depth of earth was in this way thrown off, when the large
ends of two eggs met my anxious gaze; both these eggs were
resting on their smaller apex, and the earth round them had
to be very carefully removed to avoid breaking the shell, which
is extremely fragile when first exposed to the atmosphere.
About a hundred yards from this first mound we came upon a
second, rather larger, of the same external form and appearance; it contained three eggs. Although we saw seven or
eight more mounds, only these two contained eggs; we were
too early; a week later, and we should doubtless have found
many more. . . . In both the nests with eggs the White
Ant was very numerous, making its little covered galleries of

earth around and attached to the shell, thus showing a beautiful provision of Nature in preparing the necessary tender food
for the young bird on its emergence; one of the eggs I have
preserved shows the White Ants' tracks most plainly.   The
largest mound I saw, which appeared as if in a state of preparation for eggs, measured forty-five feet in circumference, and
if rounded in proportion on the top, would have been full five
feet in height.   I remarked that, in all the mounds not ready
for the reception of eggs, the inside or vegetable portion was
always wet and cold, and I imagine, from the state of others,
that the bird turns out the whole of the materials to dry before
depositing its eggs and covering them up with the soil.   In both
cases where I found eggs, the upper part of the mound was
perfectly and smoothly rounded over, so that anyone passing
it without knowing the singular habit of the bird might very
readily suppose it to be an ant-hill; mounds in this state
always contain eggs within, while those without eggs are not
only *not* rounded over, but have the centres so scooped out
that they form a hollow.   The eggs are deposited in a very
different manner from those of the *Megapodius;* instead of each
being placed in a separate excavation in different parts of the
mound, they are laid directly in the centre, all at the same
depth, separated only by about three inches of earth, and so
placed as to form a circle."

Eggs.—When fresh, of a delicate pinky-white, but after remaining in the mound a few days, they become dirty reddish-
brown.   Shell very thin.   Average measurements, 3·5 by 2·3
inches.

THE BRUSH-TURKEYS. GENUS TALEGALLUS.

Talegallus, Lesson, Voy. " Coquille," i. pt. ii. p. 715 (1828).

Type, *T. cuvieri,* Less.

Upper tail-coverts black, and *not* extending to the end of the
tail.

Top of the head covered with *narrow* (sometimes hair-like)

feathers; sides of the head, throat, and fore-part of the neck *mostly naked. No wattle* at the base of the neck.

Bill *stout and strong.*

Nostrils *oval.*

Tail rather long, rounded, and composed of *sixteen* feathers, the *middle pair* being the longest.

First primary flight-feather *shorter than* the tenth; fifth and sixth slightly the longest.

Legs (metatarsi) and feet large; the fore-part of the legs covered with a single row of large scales.

Claws *shorter, and more rounded* than in *Megapodius* and *Eulipoa*, the claw of the middle toe being *shorter than* the upper mandible, measured from the posterior wall of the nostril to the tip.

I. CUVIER'S BRUSH-TURKEY. TALEGALLUS CUVIERI.

Talegallus cuvieri, Lesson, Voy. "Coquille," Zool. Atl. pl. 38 (1826); id. Voy. "Coquille," Zool. i. pt. ii. p. 716 (1828); Ogilvie-Grant, Cat. B. Brit. Mus. xxii. p. 465 (1893).

Adult Male and Female.—General colour black; head and back of the neck thinly covered with *narrow, almost hair-like,* feathers, which are *recumbent* on the crown; throat sparsely covered with brownish-white feathers. Bill *orange-red (yellow* in dried specimens); naked skin on the sides of the head, and neck reddish-brown; legs and feet orange or yellow. Total length, 20–21 inches; wing, 10·8; tail, 6·3; tarsus, 3·3–3·5.

Younger examples have the back and sides of the neck mostly dark chestnut.

Range.—Western New Guinea, and the Islands of Salawatti, Mysol, and Halmahèra.

Habits.—Von Rosenberg states that this species is not to be met with on the mountains, its place there being taken by *Aepypodius arfakianus.* Nothing further has been published regarding its habits.

II. THE DARK-BILLED BRUSH-TURKEY.　TALEGALLUS FUSCI-
ROSTRIS.

Talegallus fuscirostris, Salvadori, Ann. Mus. Civ. Genov. ix. pp.
332, 334 (1877); Ogilvie-Grant, Cat. B. Brit. Mus. xxii.
p. 466 (1893).

Adult Male and Female.—Like *T. cuvieri*, but the bill is *sooty-brown* instead of orange-red, and the naked skin on the sides
of the head and neck is *blackish-grey*, not *reddish*-brown. Total
length, 21 inches; wing, 11–11·5; tail, 6·8–7; tarsus, 3·5–
3·7.

Range.—Southern New Guinea, extending north-eastwards
to Geelvink Bay, and south to the Aru Islands.

Habits.—To Von Rosenberg we owe the only account we
have been able to find of the habits of this bird. He says :—
" The ' Kamur ' is not really rarer than the ' Djangul ' (*Megapodius duperreyi*), but is not met with so frequently, owing to
its solitary forest-haunting habits. Near Wonumbai I found a
new nesting-mound of this bird situated in a ' radura,' and
protected by the shade of a *Titie* (*Vitex moluccana*). It was
composed of earth mixed with sticks and leaves, the whole
forming a truncate cone 11 feet high and 25 feet round the
base. In the summit of the cone we found the openings of five
burrows which went down perpendicularly to a depth of four
feet, and were filled with earth. In four of these I found eggs
which were placed vertically. As they were broken by the
man who carried them, I was able to ascertain that they were
in various stages of development, and I was thus able to verify
the statement previously made to me by the natives, who affirm
that the eggs are laid at intervals of one or more days. In the
mound the thermometer rose to 93° Fahr., while the surrounding atmosphere was only 85° in the shade. A few days later
I found a second nesting-mound which, though it appeared to
have been abandoned for a long time, was much larger than
the first, and I was assured by my native guide that it was the

work of the female parent. The eggs are as large as those of the 'Oca,' and oblong in form; they have a hard white shell covered with a layer of reddish-brown, and are good to eat."

Eggs.—Elliptical ovals; vinaceous cinnamon in colour. Average measurements, 3·88 by 2·4 inches. (*A. B. Meyer.*)

III. THE JOBI ISLAND BRUSH-TURKEY. TALEGALLUS JOBIENSIS.

Talegallus jobiensis, Meyer, SB. Ak. Wien. lxix. Abth. i. pp. 74, 87 (1874); Ogilvie-Grant, Cat. B. Brit. Mus. xxii. p. 467 (1893).

Adult.—Plumage black as in the other species, but easily recognised by the *semi-erect crest* composed of wider and thicker feathers, which cover the top of the head. Bill and naked skin of the head dusky-red, sides of the throat blood-red, legs and feet fiery-red. "Total length, 21·5 inches; wing, 11·8; tail, 6·6; tarsus, 3·6." (*Salvadori.*)

In a nearly adult specimen in the British Museum the tail measures 7·3 inches. This bird was obtained by Hunstein in South-eastern New Guinea, opposite China Straits.

Range.—Jobi Island and the Eastern shore of Geelvink Bay; also South-east New Guinea.

Eggs.—Reddish-fawn colour; rather pointed; shell smooth. Measurements, 3·8–3·93 by 2·41–2·46 inches. (*A. B. Meyer.*)

IV. THE LONG-TAILED BRUSH-TURKEY. TALEGALLUS LONGICAUDUS.

Tallegallus longicaudus, Meyer, Abh. Mus. Dresd. 1890–91, No. 4, p. 15 (1892); Ogilvie-Grant, Cat. B. Brit. Mus. xxii. p. 467 (1893).

Said by Dr. Meyer to differ from *T. jobiensis* in being darker in colour, and in having a longer tail and shorter tarsus. Total length, 23·2 inches; wing, 11·2; tail, 8; tarsus, 3.

It seems doubtful if this bird is really distinct from *T. jobiensis*.

Range.—Astrolabe Bay, North-east New Guinea.

THE WATTLED BRUSH-TURKEYS. GENUS CATHETURUS.

Catheturus, Swains. Class. B. ii. p. 206 (1837).

Type, *C. lathami* (Lath.).

Differs from *Talegallus* in having *the head* as well as the neck almost naked, with only a few hair-like feathers; and a *large vascular wattle at the base of the neck.*

Nostrils *round*.

Tail composed of *eighteen* feathers, *the fifth pair* being considerably longer than the middle pair and much longer than the outer pair.

The fore part of the legs (metatarsi) covered with a *double row* of *hexagonal* plates.

I. THE AUSTRALIAN BRUSH-TURKEY. CATHETURUS LATHAMI.

Alectura lathami, Latham, Gen. Hist. B. x. p. 455 (1824); Jardine and Selby, Ill. Orn. iii. pl. 140.

Meleagris lindesayii, Jameson, Mem. Wernerian, Nat. Hist. Soc. vii. p. 473 (1835).

Catheturus australis, Swainson, Class. B. i. p. 284, fig. 92 (1836); ii. p. 206 (1837).

Talegalla lathami, Gould, B. Austr. v. pl. 77 (1840); Sclater in Wolf's Zool. Sketches, ii. pl. xl. (1861); North, Nests and Eggs Austr. B. p. 279 (1889).

Catheturus lathami, Reichenb. Tauben. p. 10, pl. 277, fig. 1540 (1862); Ogilvie-Grant, Cat. B. Brit. Mus. xxii. p. 468 (1893).

Adult Male and Female.—General colour of upper parts dark brownish-black, lighter on the lower back and rump; underparts dark brownish-grey, more or less conspicuously edged with white; in some examples the belly and thighs are nearly white. Naked skin of head and neck pinkish-red, the wattle bright yellow; bill black; legs brown. Total length, 25.5 inches; wing, 12; tail, 9.8–10.2; tarsus, 4–4.2.

Range.—North-east and East Australia.

Habits.—Dr. Ramsay writes:—"However plentiful this species may have been formerly in the Rockingham Bay district, it is now very scarce, only one having been obtained during my visit. They are still plentiful in the New South Wales scrubs. I found that two or more females visited the same mound to lay their eggs in, and when this is the case, the mound is often twice as large as usual. It seems probable that several individuals assist in scratching the mound together, when a space often fifty yards in diameter (on level ground) is cleared of almost every fallen leaf and twig. The mounds are often six feet in height, and twelve to fourteen yards wide at the base : sometimes they are more conical. The central portion consists of decayed leaves mixed with fine *débris*, the next of coarser and less rotten materials ; and the outside is a mass of recently gathered leaves, sticks, and twigs not showing signs of decay. In opening the nest these are easily removed and must be carefully pushed backwards over the sides, beginning at the top. Having cleared these, and obtained plenty of room, remove the semi-decayed strata, and below it, where the fermentation has begun, in a mass of light fine leaf-mould, will be found the eggs placed with the *thin end downwards*, often in a circle, with three or four in the centre, about six inches apart. At one side, where the eggs have been first laid, they will probably be found more or less incubated, but in the centre, where the eggs are placed last, quite fresh ; and if only one pair of birds have laid in the mound, about twelve to eighteen eggs will be the complement, and will be found arranged as described above. On the other hand, if several females resort to the same nest the regularity will be greatly interfered with, and two or three eggs in different stages of development will be found close to one another, some quite fresh, others within a few days of being hatched. There are usually ten eggs in the first layer, five or six in the second, three or four only in the centre. I found that the females returned every second

12

day to lay, but never succeeded in ascertaining which of the
parent birds opens the nest. The aborigines inform me that
the male bird always performs this office; and I usually found
my black boys very correct in their statements of this kind.
After robbing a nest, it is necessary to replace the different
layers as they were found; for if the lowest is too much mixed
up with the others, or the top tumbled into the excavations
made in the bottom one, the birds will invariably forsake the
mound; so that I found it always necessary to carefully replace the different layers as I found them. It is not so with
the *Megapodius duperreyi*, which species does not seem to care
how much the mound is tumbled about, so that there is sufficient *débris* left to burrow in. . . . The greatest number
of eggs taken from one mound at one time was thirty-six.
This was a very old mound, and resorted to by several individuals."

Mr. Gould observes:—"When disturbed, the Wattled Talegallus readily eludes pursuit by the facility with which it runs
through the tangled brush. If hard pressed, or when rushed upon
by its great enemy the native dog, it springs upon the lowermost
bough of some neighbouring tree, and by a succession of leaps
from branch to branch ascends to the top, and either perches
there or flies off to another part of the brush. It is also in the
habit of resorting to the branches of trees as a shelter from the
mid-day sun—a peculiarity that greatly tends to their destruction; for, like the Ruffed Grouse of America, when assembled
in small companies they will allow a succession of shots to be
fired until they are all brought down. . . .

"While stalking about the woods the Talegallus frequently
utters a rather loud clucking noise; but whether this sound is
uttered by the female only I could not ascertain; still I think
such is the case, and that the spiteful male, who appears to delight in expanding his richly-coloured fleshy wattles and un-
mercifully thrashing his helpmate, is generally mute.

"In various parts of the brush I observed depressions in

the earth, which the natives informed me were made by the birds in dusting themselves."

Eggs.—Pure white, varying much in shape and size, some being almost round, others a long oval or pointed at the smaller end ; their usual form is an oval, slightly smaller at one end. Shell thin, smooth, and minutely granulated. Average measurements, 3·55 by 2·4 inches.

THE PAPUAN WATTLED BRUSH-TURKEYS. GENUS AEPYPODIUS.

Aepypodius, Oustalet, Le Nat. No. 41, p. 323 (1880).

Type, *A. bruijni*, Oustalet.

Upper tail-coverts short and *dark chestnut.*

Head and neck mostly naked ; a *pendulous wattle* at the base of the fore-neck (in one species an additional wattle on each side of the neck) ; and an elevated *fleshy crest* extending from the base of the bill to the crown.

Nostrils *round.*

Tail composed of *sixteen* feathers and similar in shape to that of *Catheturus.*

First primary flight-feather about equal to the tenth ; seventh or eighth slightly the longest.

Front of legs (metatarsi) covered with a *single row* of large scales, the last two or three being divided down the middle.

I. THE WAIGIOU WATTLED BRUSH-TURKEY. AEPYPODIUS BRUIJNI.

Talegallus bruijnii, Oustalet, C. R. xc. p. 906 (1880).
Aepypodius bruijnii, Oustalet, Le Nat. No. 41, p. 323 (1880) ;
    Ogilvie-Grant, Cat. B. Brit. Mus. xxii. p. 470 (1893).
Talegallus (*Aepypodius*) *bruijnii*, Oustalet, Ann. Sci. Nat. xi
    p. 38, figs. 33 [adult], 34 [juv.] (1881).

Adult.—General colour above brownish-black, browner on the lower back, and dark chestnut on the upper tail-coverts ;

chest and upper breast *mostly chestnut;* rest of under-parts mostly dirty grey. In addition to the fleshy crest, the top of the head is entirely covered with *close-set horny papilli.* In addition to the wattle at the base of the fore-neck, there are a *pair of elongate wattles, one on each side of the nape.* Naked skin and wattles apparently red or orange. Total length, 19 inches; wing, 11·5–12·4; tail, 5·7–6·4; tarsus 3·8–4·2.

Range.—Island of Waigiou.

II. THE NEW GUINEA WATTLED BRUSH-TURKEY.   AEPYPODIUS
ARFAKIANUS.

Talegallus arfakianus, Salvad. Ann. Mus. Civ. Genov. ix. pp. 333, 334 (1877).

Talegallus pyrrhopygius, Schl. Notes Leyd. Mus. i. p. 159 (1879).

Talegallus (Aepypodius) pyrrhopygius, Oustalet, Ann. Sci. Nat. xi. p. 40, fig. 35 (1881).

Aepypodius arfakianus, Salvad. Ann. Mus. Civ. Genov. xviii. p. 8 (1882); Ogilvie-Grant, Cat. B. Brit. Mus. xxii. p. 470 (1893).

Adult.—May be easily recognised by the following characters :—the *back of the head* and *nape thickly covered with* black feathers; the top of the head *devoid* of papilli; the lateral wattles, so conspicuous in *A. bruijni,* absent, and the chest *brownish-black.* Total length, 17·5 inches; wing, 10·5 ; tail, 5·5 ; tarsus, 3·6.

Range.—South-east New Guinea to the West coast of Geel vink Bay.

Habits.—Nothing has been recorded, but they probably resemble the Brush-Turkeys (*Talegallus*) in their mode of life. A nesting-mound found by Beccari in the Arfak Mountains at an altitude of 6,000 feet, was probably the work of this species.

MALEO.

THE MALEOS. GENUS MEGACEPHALON.

Megacephalon, Temm.; Gray, Gen. B. iii. p. 489 (1846).

Type, *M. maleo*, Hartl.

Upper tail-coverts *much shorter* than the tail-feathers.

Head naked, covered by a *large gourd-shaped helmet;* a rounded tubercle behind each nostril ; neck and throat thickly covered with hair-like feathers.

Nostril a rather large *rounded oval.*

Tail composed of *eighteen* feathers and shaped as in *Catheturus* and *Aepypodius.*

Wing as in *Aepypodius.*

Legs (metatarsi) and feet rather long, the former covered in front with *small hexagonal scales.*

Only one species is known.

I. THE MALEO. MEGACEPHALON MALEO.

Maleo, Temm. Pl. Col. V., in text to Pl. 46 [No. 411] (1826).
Megacephalon maleo, Hartl. Verzeichniss, p. 101 (1844) ;
Ogilvie-Grant, Cat. B. Brit. Mus. xxii. p. 472 (1893).

(Plate XXXVI.)

Adult Male and Female.—Helmet black ; naked skin round eye yellowish flesh-colour ; upper-parts, chest, flanks, thighs, and under tail-coverts dark brown ; breast and belly beautiful salmon-pink ; legs bluish-black ; toes yellowish. Total length, 22 inches ; wing, 11–11·8 ; tail, 5·7–6 ; tarsus, 3·5.

In quite young examples the helmet is absent and the crown covered with mottled brown and white feathers.

Range.—Celebes and the Sanghir Islands.

Habits.—Dr. A. R. Wallace writes :—"This interesting bird is confined, so far as I am aware, to the Northern Peninsula of Celebes, and to the littoral portions of the island, never being found in the mountain ranges or in the elevated district of Tondano. It seems particularly to abound in the forests

around the base of the Klabat mountain, feeding entirely on
fallen fruits, which in the crop resemble the cotyledons of
leguminous seeds.   In the months of August and September,
when there is little or no rain, they descend to the sea-beach
to deposit their eggs.   They choose for this purpose certain
bays remote from human habitations.   One of these serves
for an extensive tract of country, and to it the birds repair
daily by scores and hundreds.   I visited the most celebrated
of these beaches, but, it being late in the season, did not see so
much of the birds as I might otherwise have done.   I made,
however, some interesting observations, and obtained a very
fine series of specimens during my stay of six days.

   " The place is situated in a bay between the island of Limbe
and Banca, and consists of a steep beach about a mile in
length, of very deep, loose, and coarse black volcanic sand
or rather gravel, exceedingly fatiguing to walk over.   It is
bounded at each extremity by a small river with hilly ground
beyond, while the forest behind the beach itself is somewhat
flat and its growth stunted, so that it has quite the appearance
of being formed from the *débris* of an ancient lava-stream from
the Klabat Volcano, especially as beyond the two rivers the
beaches are of *white* sand.   In the mass of loose sand thrown
up above high-water mark are seen numbers of holes four or
five feet in diameter.   In and around these holes, at the depth
of one or two feet, the eggs of the Maleos are found.   There
are sometimes only one or two, sometimes as many as seven or
eight in one hole, but placed each at a distance of six to eight
inches from the others, and each egg laid by a separate bird.
They come down to the beach, a distance often of ten or
fifteen miles, in pairs, and, choosing either a fresh place or an
old hole, scratch alternately, throwing up a complete fountain
of sand during the operation, which I had the pleasure of
observing several times.   When a sufficient depth is reached,
the female deposits an egg and covers it up with sand, after
which the pair return to the forest.   At the end of thirteen

days (the natives assert) the same pair return, and another egg is deposited. This statement seems to have been handed down by tradition, having perhaps originated from the observation of some wounded or singularly marked bird. I am inclined to think it is near the truth, because in the females I killed before they had laid, the egg completely filled up the lower cavity of the body, squeezing the intestines so that it seemed impossible for anything to pass through them, while the ovary contained eight or ten eggs about the size of small peas, which must evidently have required somewhere about the time named for their development. . . . The eggs when quite fresh are delicious eating, as delicate as a fowl's egg, but much richer, and the natives come for more than fifty miles round to search for them. After the eggs are once deposited in the sand the parent birds pay no further attention to them. The young birds on breaking the shell work their way up through the sand and run off to the forest.

"The appearance of the birds when walking on the beach is very handsome. The glossy black and rosy-white of the plumage, the helmeted head and the elevated tail, roofed like that of the common hen, form a *tout ensemble* quite unique, which their stately and somewhat sedate walk renders still more remarkable. When approached they run pretty quickly, and if suddenly disturbed, take flight to the lower branches of some adjacent tree. There is hardly any difference between the sexes.

"When we consider the great distances the birds come, and the trouble they take to place the eggs in a proper situation, it does seem extraordinary that they should take no further care about them. It is, however, quite certain that they neither do nor can watch over them. The eggs deposited by a number of hens in succession in the same hole must render it impossible for each to distinguish its own ; and the food of the parent birds can be obtained only by continual roaming, so that if the numbers which come down to the beach alone in the breeding season (according to the accounts, many hundreds, or even thousands) were obliged to remain in the vicinity, the greater part would perish of hunger.

"In the structure of the feet of the *Megacephalon* we may see a reason why it departs from the habits of its nearest allies, the *Megapodii* and *Talegalli*, which generally heap up mounds of earth and rubbish in which to bury their eggs. The feet of the Maleos are not nearly so strong in proportion as those of the former birds, while the claws are short and straight, instead of being very long and greatly curved."

Eggs.—Pale brownish-red. Measurements, 4·3 by 2·4 inches.

THE CURASSOWS. FAMILY CRACIDÆ.

As in the *Megapodiidæ*, the hind-toe or hallux *is on the same level* as the other toes, and its basal phalanx is as long as that of the third toe.

They differ from the *Megapodes* in having the oil-gland *tufted*.

The nest is made either in a tree or on the ground, and the eggs, which are white, are incubated in the usual manner.

THE TRUE CURASSOWS. GENUS CRAX.

Crax, Linn. Syst. Nat. i. p. 269 (1766).

Type, *C. alector* (L.).

Bill stout, the depth of the upper mandible being *greater* than the width.

Feathers on the top of the head *semi-erect and curled* at the extremity.

With or without a swollen knob at the base of the upper mandible.

Wattles at the base of the lower mandible present or absent.

Tail composed of *twelve* feathers.

In this genus the females differ one from another in plumage far more than the males, all of which are very similar in plumage.

I. THE CRESTED CURASSOW. CRAX ALECTOR.

Crax alector, Linn. S. N. i. p. 269 (1766) ; Sclater, Trans. Z. S. ix. p. 277, pl. xliii. (1875) ; Ogilvie-Grant, Cat. B. Brit. Mus. xxii. p. 475 (1893).

Crax mitre (*nec* L.), Vieillot, Gal. Ois. ii. pl. 199 (1825).

Crax sloanei, Reichenb. Tauben, p. 131 (1862).

Crax erythrognatha, Sclater and Salvin, P. Z. S. 1877, p. 22 ; Sclater, Trans. Z. S. x. p. 543, pl. xc. (1879).

Adult Male.—General colour black, *glossed with purple*, except the belly, flanks, and under tail-coverts, which are *white;* crest *uniform black;* tail *not* tipped with white ; *no* swollen knob at the base of the upper mandible nor wattles on the sides of the lower ; cere and base of bill yellow, the tip horny blue. Total length, 34 inches ; wing, 15 ; tail, 13 ; tarsus, 4·5.

Adult Female.—Like the male, but the feathers of the crest with a few white bars. Rather smaller; wing, 14 inches.

Range.—Northern South America. Rio Negro, Rio Branco, Rio Vaupè, British Guiana, and the United States of Colombia.

Habits.—According to Sonnini this species is very numerous in French Guiana and met with in large flocks in the vast forests which cover the greater part of that country. He found it of a remarkably tame and confiding disposition, and by no means afraid of his presence, but in the more inhabited parts it was much wilder. It appears to nest during the rainy season, which in Guiana lasts for seven or eight months.

The "Mituporanga" generally keeps to the mountain forests, perching on the high trees; and passes much of its time on the ground searching for fruits, which form its chief diet. Tame examples of this species are frequently to be seen in the streets of the town of Cayenne, and may be observed entering all the houses in the most fearless manner, and searching beneath the tables for food.

Mr. C. B. Brown met with large numbers of this species along the banks of the River Corentyne, and was able to

shoot as many as he required for food as his boat was passing
along.

Nests.—Placed in a tree; composed of sticks, and coarsely
lined with dry grass and leaves.

Eggs.—Pure white; two to six in number.

II. SCLATER'S CURASSOW. CRAX FASCIOLATA.

Crax fasciolata, Spix, Av. Bras. ii. p. 48, pl. lxii. *a* (1825);
    Ogilvie-Grant, Cat. B. Brit. Mus. xxii. p. 476 (1893).
Crax sclateri, Gray, List Gallinæ Brit. Mus p. 14 (1867)
    [part].

Adult Male.—Like the male of *C. alector*, but the plumage is
black, glossed with *dark green*, and the *tips of the tail-feathers
are white*. Total length, 30·5 inches; wing, 14·3; tail, 13·6;
tarsus, 4.

Adult Female.—Crest white, with the base and tip of each
feather black; upper-parts and tail black, with *narrow white
cross-bars*, widest on the wings;* outer primary quills *black,
barred with white;* chest, and sometimes the sides of breast,
buff, barred with black; thighs and rest of under-parts pale
rufous-buff. Smaller than the male. Wing, 13·2 inches.

Range.—Forests of Eastern South America, extending north
to Para, south to Paraguay, and thence east to Bolivia.

III. NATTERER'S CURASSOW. CRAX PINIMA.

Crax pinima, Natterer, MSS.; Pelz. Orn. Bras. pp. 287, 341
    (1870); Ogilvie-Grant, Cat. B. Brit. Mus. xxii. p. 477
    (1893).
Crax incommoda, Sclater, P. Z. S. 1872, p. 690; id. Trans.
    Z. S. ix. p. 281, pl. xlix. (1875); x. p. 544, pl. xciii.
    (1879).

Adult Male.—At present unknown.

---

* These white bars appear to decrease with age; in one of the examples
examined they have entirely disappeared on the tail.

Adult Female.—Differs from the female of *C . fasciolata* in having the tail-feathers uniform black (the *middle pair* only being irregularly marked with white in a younger example); the *thighs*, as well as the breast and sides, *black, barred with pale buff;* and, in the adult, at least, the crest is black, narrowly barred with white. Total length, 34'5 inches; wing, 15'2 ; tail, 13'5; tarsus, 4'1.

Range.—South America. District of Para, perhaps to the United States of Colombia.

IV.   THE MEXICAN CURASSOW.   CRAX GLOBICERA.

Crax globicera, Linn. S. N. i. p. 270 (1766); Sclater, Trans. Z. S. ix. p. 274, pl. xl. (1875) [part]; Ogilvie-Grant, Cat. B. Brit. Mus. xxii. p. 478 (1893).
Crax rubra, Linn. S. N. i. p. 270 (1766).
Crax temminckii, Tschudi, Faun. Per. p. 287 (1844–46).
Crax pseudalector, Reichenb. Tauben, p. 131, pl. 174, fig 15-16, and *C. edwardsi*, p. 134 (1862).

Adult Male.—Black, glossed with dark green, except the middle of the belly, flanks, and under tail-coverts, which are white ; *a swollen yellow knob* at the base of the upper bill. Total length, 35 inches ; wing, 15'7 ; tail, 14 ; tarsus, 4'6.

Adult Female.—Crest feathers widely *barred with white* across the middle; rest of head, neck, and throat barred with black and white ; mantle and chest black, washed or margined with rufous, and glossed with green; lower back mostly deep brownish-chestnut ; quills and wing-coverts *chestnut*, mottled with black ; breast deep chestnut, shading into cinnamon on the rest of the under-parts; *tail* black, the median feathers generally mottled with chestnut, and with traces of irregular yellowish-white bars. Size smaller ; wing, 14 inches.

Range.—Central America, extending from Western Mexico to Honduras and Cozumel Island.

204 LLOYD'S NATURAL HISTORY.

Habits.—Mr. G. F. Gaumer, writing from Yucatan, says :—
"This is a very shy bird, living far in the interior of unin-
habited forests. Its walk is cautious and almost noiseless ; it
is generally found in pairs, though the males often travel alone.
It spends most of its time upon the ground, where it finds its
food by scratching among the leaves. In the morning and
evening it mounts upon the trees which bear its favourite food,
to feast upon the best fruits of the forest. It ascends not by
a single flight, but by short flights from limb to limb, until it
reaches the fruit. While there it makes no noise, but at every
moment it listens for the approach of an enemy, which once
discovered, it utters a short, impatient cluck, and flies away to
a very great distance. The song resembles the deep distant
bass roaring of the tiger, or the gentle blowing in the bung-
hole of a barrel. The flesh of this bird is highly valued as
food, but the bones are always carefully kept away from the
dogs and cats, as they are said to be very poisonous. It is
sometimes domesticated, though it rarely lives beyond a few
months."

Mr. Charles W. Richmond says :—"This bird is rather
common. Observed on the Rio Frio, and on the Escondido.
It is often kept in captivity. A fine male on the Magnolia
plantation was very tame, and answered to the name of
"Touie." One of Touie's peculiarities was an abhorrence of
women. The moment a dress appeared on the plantation
he began to show great distress, uttering his low, plaintive
whistle, and running after the object of his wrath, with body
leaning forward and almost brushing the ground, head thrown
back, and tail raised, giving him a laughable appearance.
After picking at the offending dress, and following its wearer
about for a time, Touie would quiet down for a bit, but would
continue to sulk and utter his note of complaint until the
cause of the trouble had departed. This bird raised its crest
when excited, or when its curiosity was aroused, but on other
occasions kept it depressed."

V. THE PANAMA CURASSOW. CRAX PANAMENSIS.

Crax globicera, Sclater (nec Linn.), Trans. Z. S. x. p. 543, pl. lxxxix. (1879).

Crax rubra, Stephen (nec Linn.), in Shaw's Gen. Zool. xi. p. 168, pl. ix. (1819).

Crax alberti, Fraser, P. Z. S. 1850, p. 246, pls. xxvii. and xxviii. [part, female].

Crax panamensis, Ogilvie-Grant, Cat. B. Brit. Mus. xxii. p. 479 (1893).

Adult Male.—Like the male of *C. globicera*, but the tail-feathers are slightly margined with white. The colours of the soft parts may also differ, but they are at present unknown. Total length, 34·5 inches; wing, 14·6; tail, 13·3; tarsus, 4·2 inches.

Adult Female.—Easily distinguished from the female of *C. globicera* by having the tail *strongly barred with white* or pale buff, the bars being *as clearly marked on the under-surface* of the feathers as on the upper. Size smaller; wing, 14·4 inches.

Range.—Southern Nicaragua and Costa Rica, extending to the United States of Colombia.

Habits.—Dr. Von Frantzius writes: " This beautiful Hocko is called in Costa Rica ' Pajuil,' which is a corruption of the Mexican word ' Pauxi.'

" I saw it in a wild state for the first time on the Sarapiqui in December, 1853. I have often seen tame specimens since in the aviaries of the chief town of San José. The young can be easily tamed if caught and reared. This bird is often shot on account of its savoury flesh."

VI. HECK'S CURASSOW. CRAX HECKI.

Crax hecki, Reichenow, J. f. O. 1894, p. 231, pl. ii

Adult Female.—Head and *neck black, barred with white;* crest-feathers black, with two or three white bands; back, chest, wing-coverts, and secondary quills barred with buff and rufous, the rufous bars being margined with black; primaries barred with buff and dusky; belly fawn-colour, the breast indistinctly barred with rufous; tail-feathers black, with narrow buff bands. Wing, 16 inches; tail, 14; tarsus, 4. (*Reichenow.*)

Range.—Unknown.

This species, recently described by Dr. Reichenow from a female specimen, appears to belong to a form intermediate between *C. panamensis* and *C. grayi.* It seems to resemble the latter in having the back barred, but differs from both in having the neck banded with black and white. Dr. Reichenow says that during the two years this bird lived in the Zoological Gardens at Berlin, the plumage underwent marked changes, the white bars becoming more pronounced with age.

VII. GRAY'S CURASSOW.   CRAX GRAYI.

Crax grayi, Ogilvie-Grant, Cat. B. Brit. Mus. xxii. p. 480 (1893).

(*Plate XXXVII.*)

Adult Male.—Unknown.

Adult Female.—Easily distinguished from the female of *C. fasciolata* and those previously described by having the primary and secondary quills, as well as the tail-feathers, *widely barred with black and white*, the white bars being about as wide as the black interspaces; the lower back, rump, and upper tail-coverts tawny buff, indistinctly barred with black. Total length, 32 inches; wing, 13·6; tail, 12·7; tarsus, 3·8.

Range.—South America. The exact locality is as yet unknown.

VIII. THE WATTLED CURASSOW.   CRAX CARNUCULATA.

Crax carnuculata, Temm. Pig. et Gall. iii. pp. 44, 690 (1815); Ogilvie-Grant, Cat. B. Brit. Mus. xxii. p. 481 (1893).

Crax rubrirostris, Spix. Av. Bras. ii. p. 51, pl. lxvii. (1825).

Crax yarrellii, Jardine and Selby, Ill. Orn. n. s. pl. vi. (1836).

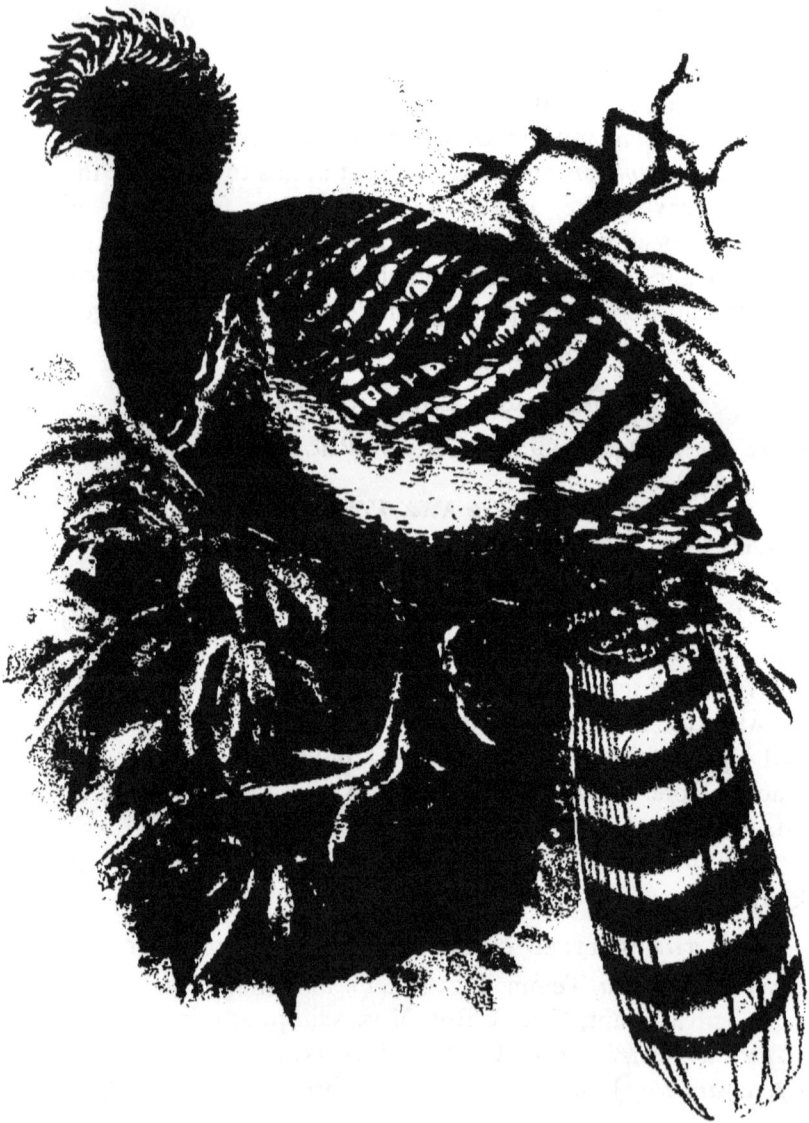

PLATE XXXVII.

GRAY'S CURASSOW.

Crax blumenbachii, Burmeister, Syst. Uebers. iii. p. 345 (1856).

Adult Male.—Plumage like that of *C. globicera,* the tail *not* tipped with white; the large *swollen knob* at the base of the upper mandible and the *wattle* on each side of the base of the lower mandible *scarlet.* Total length, 32 inches; wing, 14·8; tail, 13·5; tarsus, 4.

Adult Female.—Differs from the male in having the feathers of the crest indistinctly barred with white; the belly, flanks, and under tail-coverts *rufous-buff;* the swollen knob and wattles absent; and the basal half of the bill *scarlet.* Size rather smaller, wing, 14·2.

Range.—South-eastern Brazil, extending from Rio de Janeiro to Bahia.

Habits.—From Prince Maximilian of Neuwied's work we translate the following interesting account of this bird's habits.

"The 'Mutung' is a beautiful large bird which is only to be met with where it can find a safe home in secluded parts of the forest. I have often found it in such places, living in pairs, even out of the pairing season. I have not come across it farther south than the rivers Itapemirim and Itabapuana, but it is often found on the Rio Doce, Mucuri, Alcobaca, Belmonté, and is everywhere a very favourite game bird. It replaces in those forests our European Capercailzies. In the pairing time, especially in November, December, and January, the far-reaching cry of the cock is heard far and wide calling the hens round him. He then, it is said, spreads his tail, makes all kinds of movements with his wings, and calls in a deep tone hu! hu! hu! hu! which can be heard a long way off. These birds live much on the ground, and are therefore often caught in snares. They feed on fruits, for I have found hard fruits and nuts in their crops, both partly and entirely digested, and which were sometimes so hard that one could not cut them with a knife. I did not find stones, though the birds must undoubtedly swallow them. . . ."

"These beautiful birds are eagerly hunted by the Brazilians, especially at the season when their loud deep voice is heard, when it is not difficult to surprise them, as in sparsely inhabited districts they are by no means shy. Their flesh is excellent, and the large strong feathers of the wings and tail are used by the savages for their arrows. The Mutung would be very useful, if domesticated, as it is easily tamed. The inhabitants of some districts are well aware of this fact, for tame birds may often be met with amongst them. The Portuguese are not fond of keeping them in captivity, as they swallow anything bright or glittering, as, for instance, money, buttons, &c., which become quite useless through the great muscular powers of the stomach."

According to Burmeister it inhabits the wooded districts on the East Coast of Brazil, from Rio Janeiro to Bahia, and is known there under the name of 'Mutung.' He only once acquired a specimen from Rio de Pomba, and says that it is seldom found in the more thickly populated districts and is difficult to obtain. As a rule it does not perch very high up on the trees, and is found most frequently in dark copses in the underwood, making its nest either there or *quite on the ground.** Its food is mostly picked up on the ground, and consists as a rule of fallen nuts and the larger kinds of seed.

Nest.—Is said to be placed in a tree, and composed of sticks and twigs.

Eggs.—Said to be four in number, large, and whitish in colour.

IX. SPIX'S WATTLED CURASSOW.   CRAX GLOBULOSA.

Crax globulosa, Spix, Av. Bras. ii. p. 50, pls. lxv. lxvi. (1825); Sclater, Trans. Z. S. ix. p. 279, pl. xlvi. (1875); x. p. 544, pl. xci. (1879); Ogilvie-Grant, Cat. B. Brit. Mus. xxii. p. 482 (1893).

---

* This statement is probably incorrect.

Adult Male.—Only differs from the male of *C. carunculata* in having the swollen knob and wattles *yellow*, instead of scarlet.

Adult Female.—Only differs from the female of *C. carunculata* in having the basal part of both mandibles *yellow*, instead of scarlet.

Range.—Upper Amazons ; Pebas, Rio Napo, Rio Marañon, Rio Ucayali, and Sarayacu.

It seems more than probable that this species may prove to be identical with *C. carunculata ;* in dried skins of the latter the scarlet knob and wattles are always yellow ; and, so far as I am aware, living examples of *C. globulosa* have never been brought to Europe, so that Spix is the only authority for the statement that the colour of the soft parts is yellow, and he may have taken his description from dried skins. The fact that the females of both are identical in plumage is even stronger evidence that the two forms belong to the same species. Further evidence is wanted, and it is to be hoped that any naturalist or sportsman visiting the Upper Amazons will endeavour to settle this point. The colours of the soft parts should be noted as soon after death as possible, for they change rapidly.

X. DAUBENTON'S WATTLED CURASSOW.　CRAX DAUBENTONI.

Crax daubentoni, Gray, List. Gallinæ Brit. Mus. p. 15 (1867) ；
　　Sclater, Trans. Z. S. ix. p. 276, pls. xli. xlii. (1875) ；
　　Ogilvie-Grant, Cat. B. Brit. Mus. xxii. p. 482 (1893).
Crax mikani, ♂ [nec ♀], Pelz. Orn. Bras. p. 343 (1870).

Adult Male.—Distinguished from the male of *C. globicera* by having the tail-feathers *tipped with white*, and a *pale yellow wattle* on each side of the basal part of the lower mandible. Total length, 34 inches ; wing, 15·2 ; tail, 13·5 ; tarsus, 4·6.

Adult Female.—Differs from the male in having the feathers of the crest *white* near the base ; the breast, sides, and thighs *barred with white* (as in the adult female of *C. pinima*) ; the wing-coverts more or less marked with lines of the same colour, and the *secondary quills uniform black*. These characters, as well as the white belly and the tail-feathers being *tipped with white*, distinguish it from the females of all the allied species. Size smaller; wing, 14·5 inches.

Range.—Northern South America ; Venezuela.

Habits.—In 1871, Mr. A. Warmington forwarded three living examples of this species to the Gardens of the Zoological Society, and furnished the following notes :—" The three Curassows (one male and two females) were captured at 'Maron' near Tucacas, N. Venezuela, and at the present time are nearly two years old, having been taken from the nest when scarcely larger than a chick of two months old. They soon became perfectly tame and would follow me about. When able to fly they made short flights, always quickly returning and seldom alighting. At night they invariably roosted on the highest spot they could find in the home corral. They are called by the natives ' Porū.' Their cry is a sort of mournful prolonged whistle, and in the forest, when eight or ten are together, has a very singular effect. It is not common to see these birds on the ground. When they alight in a tree they almost invariably utter their cry, and at the same time raise the tail-feathers like a fan, thus exposing the white plumage beneath, and offering a conspicuous and tempting mark for the sportsman. They are excellent eating. I have never heard of these birds breeding in confinement, though I cannot say they do not. The young ones are exceedingly beautiful and delicate little creatures, marked very much like, and having a very similar appearance to young Partridges or Quails. They become much attached to individuals who treat them kindly. These birds are common in all parts of Venezuela where there is a forest."

Daubenton's Wattled Curassow is only found in the low country ; in the mountains its place appears to be taken by the Helmeted Curassow (*Pauxis pauxi*).

XI. PRINCE ALBERT'S WATTLED CURASSOW. CRAX ALBERTI.

Crax alberti, Fraser, P. Z. S. 1850, p. 246, pls. xxvii. xxviii. [nec ♀]; Sclater, Trans. Z. S. ix. p. 280, pl. xlviii. (1875); Ogilvie-Grant, Cat. B. Brit. Mus. xxii. p. 483 (1893).

Crax viridirostris, Sclater, Trans. Z. S. ix. p. 282 (1875); x. p. 544, pl. xcii. (1879).

Adult Male.—Like the male of *C. daubentoni,* but the swollen knob at the base of the bill and the wattles on each side of the base of the lower mandible are *blue,* and the lores are thickly covered with feathers. Total length, 35 inches; wing, 16; tail, 15; tarsus, 4·6.

Adult Female.—Most like the female of *C. fasciolata,* but distinguished by having the crest-feathers nearly black, with two *narrow white bars ;* the *lores densely feathered ;* the under-parts of a much deeper chestnut colour ; and the outer primary quills *chestnut.* Size smaller ; wing, 15·2 inches.

Range.—North-eastern South America; United States of Colombia.

THE FLAT-CRESTED CURASSOWS. GENUS NOTHOCRAX.

Nothocrax, Burmeister, Syst. Uebers. iii. p. 347 (1856).

Type, *N. urumutum* (Spix).

The height of the upper mandible is *greater than* the width.

A full crest of *long recumbent* feathers covers the top of the head.

Lores and a large space around the eye naked.

Tail composed of *twelve* feathers.

Leg (metatarsus) longer than middle toe and claw.

Only one species is known.

I. THE FLAT-CRESTED CURASSOW. NOTHOCRAX URUMUTUM.

Crax urumutum, Spix, Av. Bras. ii. p. 49, pl. lxii. (1825).

Urax urumutum, Burmeister, Syst. Uebers. iii. p. 347 (1856).
Nothocrax urumutum, Burmeister, Syst. Uebers. iii. p. 347
(1856); Sclater, Trans. Z. S. ix. p 282, pl. 50; x. p. 545,
pl. xciv. (1879); Ogilvie-Grant, Cat. B. Brit. Mus. xxii.
p. 484 (1893).

Adult Male.—Crest black; feathered parts of head, throat,
neck, and chest chestnut, shading into brownish chestnut on
the upper-parts, and all finely mottled with black; breast and
rest of under-parts cinnamon, with some dusky mottling on the
sides; outer webs of the secondary quills mottled with rufous-
buff; tail black, tipped with whitish-buff. Naked space round
eye yellow above, purplish below; bill scarlet; legs flesh-
colour. Total length, 24 inches; wing, 11·5; tail, 9; tarsus,
3·5; middle toe and claw, 2·9.

Adult Female.—Differs from the male in having the upper-
parts and middle tail-feathers more coarsely mottled with pale
rufous-buff on a darker ground; and the chest, breast, and
sides clouded with dusky. Size smaller; wing, 10·5 inches.

Range.—British Guiana, Rio Negro, and Upper Amazons to
Rio Pastaza and Sarayacu in Ecuador.

Habits.—According to Natterer, this bird lives during the
day in hollow trees or the thickest part of the woods, and is
very seldom met with by sportsmen; but when found it
behaves with extreme stupidity, and is caught by the Indians
with a loop fastened to the end of a pole. It searches for food
during the night, and its cry is heard before midnight and daybreak. The Indians light torches, and follow the cry till they
are near the bird, when they extinguish the light, and wait for
daybreak to kill it.

Mr. E. Bartlett writes:—"I first saw this beautiful species of
Curassow in a Peruvian's house, at Santa Maria on the Huallaga,
where it was running about along with the common fowls. The
bird appeared to be lively and active, and would fight the dogs
and fowls, driving them out of the house. A very curious cir-

cumstance is that when one of the hens commenced sitting the bird would drive her off the nest and take her place ; this I witnessed myself; the attempt at incubation, however, was not of long duration, for the Curassow destroyed the eggs, as I was informed afterwards by the owner.

"I ascertained that the bird came from the Rio Pastaza ; and I believe it is not uncommon on that river and thoughout the dense forests on the north-west bank of the Amazons.

"I have often heard this bird in the middle of the night near Nauta.

"The Peruvians called it the 'Monte Píyu.'

"The habits of this bird render it most difficult to obtain, from its living in holes or burrows in the ground. The Indians remain in the forest all night at the place where it is heard. I was informed by the Peruvians, whose word I could rely upon, that these birds come out at night, and ascend to the top branches of the lofty trees in search of food. The Indians are on the look-out, and shoot them just before sunrise as they are descending to return to their places of concealment, where they pass the day."

THE RAZOR-BILLED CURASSOWS. GENUS MITUA.

Mitu, Lesson, Traité d'Orn. p. 485 (1831).
Mitua, Strickl. Ann. Mag. N. H. vii. p. 36 (1841).

Type, *M. mitu* (Linn.).

Upper mandible *much elevated*, the height being *greater than* the width.

Crest moderate or well-developed, not curled at the extremity.

Lores thickly feathered.

Tail composed of twelve feathers.

Leg (metatarsus) longer than the middle toe and claw.

Sexes similar in plumage.

I. THE RAZOR-BILLED CURASSOW.   MITUA MITU.

Crax mitu, Linn. S. N. i. p. 270 (1766).
Pauxi mitu, Temm. Pig. et Gall. iii. pp. 8, 685 (1815).
Ourax mitu, Cuv. Règne An. i. p. 441 (1817); Temm. Pl. Col.
   v. pl. 20 [no. 153] (1823).
Crax tuberosa, Spix, Av. Bras. ii. p. 51, pl. lxvii. *a* (1825).
Ourax erythrorhynchus, Swains. Class. B. ii. p. 352 (1837).
Mitua brasiliensis, Reichenb. Tauben, p. 137 (1862).
Mitua tuberosa, Sclater, Trans. Z. S. ix. p. 283, pl. li. (1875).
Mitua mitu, Ogilvie-Grant, Cat. B. Brit. Mus. xxii. p. 486
   (1893).

Adult Male and Female.—General colour black, glossed with blue; belly and under tail-coverts dark chestnut; *tail tipped with white.* Crest well developed; upper mandible swollen and elevated.

Male: Total length, 34 inches; wing, 14·5; tail, 12; tarsus, 4·4; middle toe and claw, 3·5.

Female: Somewhat smaller.

Range.—British Guiana, extending eastwards to Para, south along the Rio Tapajos and Rio Madeira to Matto Grosso, also to Bolivia, westwards to Peru, the Rio Marañon, and the Upper Amazons.

Mr. H. W. Bates, who met with numbers of this species on the Rio Tapajos, writes:—"We were amused at the excessive and almost absurd tameness of a fine Mutum or Curassow Turkey that ran about the house. It was a large glossy-black species, having an orange-coloured beak, surmounted by a bean-shaped excrescence of the same hue. It seemed to consider itself as one of the family, attended at all the meals, passing from one person to another round the mat to be fed, and rubbing the sides of its head in a coaxing way against their cheeks or shoulders. At night it went to roost on a chest in a sleeping-room beside the hammock of one of the little girls, to whom it seemed particulary attached, following her wherever she went about the grounds. I found this kind of Curassow bird was very common in the forests of the Cupári; but it is rare on the Upper Amazons. These birds in their natural state never descend from the tops of the loftiest trees, where they live in small flocks and build their nests. It is difficult to find the reason why these superb birds have not been reduced to domestication by the Indians, seeing that they so readily become tame. The obstacle offered by their not breeding in confinement, which is probably owing to their arboreal habits, might, perhaps, be overcome by repeated experiment; but for this the Indians probably have not sufficient patience or intelligence."

Nest.—Built of sticks, &c., and placed in a tree.

Eggs.—Two in number, white, and rough-shelled.

II. THE LESSER RAZOR-BILLED CURASSOW. MITUA TOMENTOSA.

Crax tomentosa, Spix, Av. Bras. ii. p. 49, pl. lxiii. (1825).
Pauxi tomentosa, Gray, Gen. B. iii. p. 487 (1846).
Urax tomentosa, Burmeister, Syst. Uebers. iii. p. 349 (1856).
Mitua tomentosa, Sclater, Trans. Z. S. ix. p. 280, pl. lii. (1875);
    Ogilvie-Grant, Cat. B. Brit. Mus. xxii. p. 486 (1893).

Adult Male and Female.—General colour black glossed with purplish-blue in the male, and with blue in the female; belly, under tail-coverts, outer part of thighs, and *tips of tail-feathers dark chestnut.* Crest short. Upper mandible not much swollen.

Male: Total length, 35 inches; wing, 15; tail, 13·3; tarsus, 4·8; middle toe and claw, 3·8.

Female: Somewhat smaller; wing, 14 inches.

Range.—British Guiana, extending southwards along the Rio Branco and Rio Negro.

III. SALVIN'S RAZOR-BILLED CURASSOW. MITUA SALVINI.

Mitua salvini, Reinhardt, Vid. Medd. Nat. Forh. Kjöbenhavn, Jan. 8th, 1879, pp. 1-6; id. P. Z. S. 1879, p. 108; Sclater, Trans. Z. S. x. p. 545, pl. xcv. (1879).

Adult Male and Female.—General colour black, glossed with blue; *lower part of abdomen, under tail-coverts, and tips of tail-feathers, white.* Crest elongate, as in *M. mitu.*

Male: Total length, 28 inches; wing, 14·5; tail, 12·7; tarsus, 5; middle toe and claw, 3·9.

Female: Slightly smaller.

Range.—Sarayacu, Ecuador.

THE HELMETED CURASSOWS. GENUS PAUXIS.

Pauxi, Temm. Pig. et Gall. ii. pp. 456, 465, 468 (1813); iii. pp. 1, 683 (1815).

Pauxis, Sclater, Trans. Z. S. ix. p. 285 (1875).

Type, *P. pauxi* (Linn.).

A large, elevated, *egg-shaped casque, or helmet,* covering the base of the upper mandible and forehead.

Feathers of the head and neck short and velvety.

Lores thickly feathered.

Tail composed of twelve feathers.

Leg (metatarsus) longer than the middle toe and claw

Only one species is known.

I. THE HELMETED CURASSOW. PAUXIS PAUXI.

Crax pauxi, Linn. S. N. i. p. 270 (1766); Vieill. Gal. Ois. ii. p. 5, pl. 200 (1825).

Crax galeata, Latham, Ind. Orn. ii. p. 624 (1790).

Pauxi galeata, Temm. Pig. et. Gall. iii. pp. i. 683 (1815); Gray, Gen. B. iii. p. 487, pl. cxxii. (1846)..

Pauxis pauxi, Sclater, Trans. Z. S. ix. p. 285, pl. liii. fig. i. (1875); Ogilvie-Grant, Cat. B. Brit. Mus. xxii. p. 488 (1893).

Pauxis galeata, var. *rubra*, Sclater, Trans. Z. S. ix. p. 285, pl. liii. fig. 2 (1875).

Adult Male.—General colour black, glossed with dark green;

abdomen, under tail-coverts, and tips of tail-feathers white; casque slate-blue; bill and legs red. Total length, 33·5 inches; wing, 14; tail, 13·2; tarsus, 4·2; middle toe and claw, 3·5.

Adult Female.—Differs chiefly from the male in having the chin and throat mostly pale reddish-brown; the back and wing-coverts chestnut, barred with black and tipped with buff; the lower back reddish-brown, with indistinct black bars; the chest rufous, barred with black, and the breast, sides, and flanks rufous-buff. Size similar.

Some females resemble the male in plumage. Mr. Dawson Rowley records an undoubted female in black plumage similar to that of the male. Further information is required on this subject.

Range.—Venezuela; United States of Colombia; Rio Cassiquiari, and R. Orinoco; North-east and Central Peru. Buffon records this species from Cayenne.

Mr. W. Summerhayes says that in the mountains of Venezuela this bird is common, and takes the place of *Crax daubentoni*, which is only found along the littoral as far as the foot of the mountains.

Eggs.—Broad ovals; shell white and rough. Average measurements, 3·5 by 2·5 inches.

THE AMERICAN MOUNTAIN-PHEASANTS. GENUS OREOPHASIS.

Oreophasis, G. R. Gray, Gen. B. iii. p. 485 (1844).

Type, *O. derbianus*, Gray.

Width of the upper mandible *greater than* the height.

An *elongate, straight, rather slender, cylindrical casque or helmet* situated on the top of the head between the eyes; crown mostly naked.

Base of the upper mandible, as far as the nostrils, *densely covered with velvety feathers.*

Tail composed of twelve feathers.

First primary flight-feather *much the shortest*, about half the length of the fifth, which is equal to the tenth; seventh slightly the longest.

Leg (metatarsus) slightly longer than the middle toe and claw.

Sexes similar in plumage.

Only one species is known.

I. THE EARL OF DERBY'S MOUNTAIN-PHEASANT.   OREOPHASIS
DERBIANUS.

Oreophasis derbianus, Gray, Gen. B. iii. p. 485, pl. cxxi. (1844) ;
Sclater and Salvin, Ibis, 1859, p. 224 ; Ogilvie-Grant, Cat.
B. Brit. Mus. xxii. p. 489 (1893).
Penelope fronticornis, Van der Hoev. Handb. der Zool. ii. p.
435 (1852-56).

Adult Male and Female.—General colour above black, glossed with dark green ; base of upper mandible, forehead, and sides of the head velvety black ; base of throat nearly naked ; chest and breast white, with dark shaft stripes ; sides mostly buff, with dark centres to the feathers ; a wide white band across the middle of the tail.   Helmet (thinly covered with hair-like feathers), legs, and feet, deep vermilion ; bill, pale straw-colour ; iris, white.

Male: Total length, 36 inches ; wing, 15·5 ; tail, 15·2 ; tarsus, 3·6 ; middle toe and claw, 3·4.

Female: Somewhat smaller ; wing, 14·8 inches.

Range.—Central America ; woods of the Volcan de Fuego, Guatemala.

Habits.—This fine bird, one of the most interesting to be met with in Central America, still remains one of the rarest prizes, and, so far as I am aware, is only to be found on the Volcan de Fuego. Previous to 1860, only about seven specimens were known, but in that year Mr. Osbert Salvin, under the guidance of one José Ordoñez, a hunter from Dueñas, paid a visit to the spot. José had on several previous occasions succeeded in shooting specimens of the Mountain-Pheasant, but on this expedition none were obtained. Mr. Salvin writes :— " We started at six o'clock in the morning at break of day, reached the forest region at nine, and continued climbing until we had almost passed out of it into the region of pines and coarse grass with which the peak is clothed, but no *Oreophasis* was met with. Descending again, we struck the barranco in which José had shot the specimens he brought me, but with no better success, except that I found unmistakable 'sign' in the shape of feathers, and the fruit of the tree I had been in search of. Though not successful, this expedition was satisfactory in one respect—I had seen a spot where the *Oreophasis* certainly had visited, and where my specimens had been killed. . . . I regret that I cannot give any other than José's account of the habits of this bird, but as his stories bear a semblance of truth, I do not hesitate in transcribing them. In the early morning he told me he usually found them in the upper branches of the forest trees, searching for their favourite fruit (a species of *Prunus*) which they eat both ripe and unripe ; as the day advances they descend to the underwood, where they remain all day, basking and scratching among the leaves. This is pretty much what a *Penelope* or *Crax* does, both of which I have frequently had opportunities of observing in the forests of the lowlands. The cry of the bird he could not describe satisfactorily. . . ."

The *Oreophasis* is known to the Indians frequenting the mountains as " Khannanay " and to the Ladinos, or half-breed Indians, as the " Faisan."

Mr. Salvin informs me that he subsequently shot several additional examples of the Mountain-Pheasant, but his personal observations have not yet been published and will appear at some future time in Messrs. Godman and Salvin's great work, the "Biologia Centrali Americana."

THE PENELOPES. GENUS PENELOPE.

Penelope, Merrem, Beytr. Vögel, pt. ii. p. 42 (1786); id. Av. Icones et Descr. ii. p. 39 (1786).

Type, *P. marail*, Gmel.

Width of the upper mandible *greater than* the height.

Nostrils situated rather far forward, never covered with feathers; top of the head feathered; a large naked space round the eye.

Chin and throat naked* with a median wattle.

Tail composed of twelve feathers.

The inner web of the outer primary quills not very deeply excised at the extremity.

Leg (metatarsus) longer than, or about equal to, the middle toe and claw.

Sexes similar in plumage.

Fifteen species are known; some being about the size of a half-grown Turkey, others as small as the Common Pheasant.

I. THE WHITE EYE-BROWED PENELOPE.   PENELOPE SUPERCILIARIS.

Penelope superciliaris, Illiger; Temm. Pig. et Gall. iii. pp. 72, 693 (1815); Ogilvie-Grant, Cat. B. Brit. Mus. xxii. p. 491 (1893).

Penelope jacupemba, Spix, Av. Bras. ii. p. 55, pl. lxxii. (1825).

Salpiza superciliaris, Wagler, Isis, 1832, p. 1226.

Adult Male and Female.—Easily distinguished from all other species of *Penelope* by having the wing-coverts and shoulder feathers (scapulars) *clearly bordered with light rufous.* Feathers of the crown uniform dark brown; eyebrow-stripes

---

* In *Penelope montagnii*, *P. sclateri*, and *P. argyrotis*, the chin and upper part of the throat are sometimes partially feathered.

white and well-defined; otherwise like the following species, *P. montagnii*. Total length, 24 inches; wing, 9·5; tail, 11·3; tarsus, 2·8; middle toe and claw, 2·6.

Range.—Brazil, Para, Bahia, São Paulo, Rio Grande do Sul, Rio Parana, and Matto Grosso.

Habits.—Prince Maximilian of Neuwied says:—"I have found the 'Schacupemba' in all parts of the forest I have traversed, and it is observed even in those parts directly washed by the waves of the sea. Although the bird is not scarce, it must be carefully looked for in the thick interwoven branches, when it does not immediately fly off. I have never observed the 'Schacupemba' on the ground, but always about half-way up the trees. It has a short, harsh call, frequently repeated, from which one may conclude that its trachea is not very highly developed. I found the remains of fruit and insects in its crop. The flesh is delicious. The Indians in many districts tame these birds, and they run about in the woods round their huts."

Nest.—Placed in a tree, and composed of sticks and twigs.

Eggs.—Two to four in number.

II. MONTAGNE'S PENELOPE. PENELOPE MONTAGNII.

Ortalida montagnii, Bonap. C. R. xlii. p. 875 (1856).
Penelope montagnii, Gray, List of Gallinæ, Brit. Mus. p. 8 (1867); Ogilvie-Grant, Cat. B. Brit. Mus. xxii. p. 492 (1893).
Stegnolæma montagnii, Sclater and Salvin, P. Z. S. 1870 p. 521.

Adult Male and Female.—Head, back and sides of the neck, and feathers on the chin and throat dark brown, *margined with pale grey ;* mantle, wings, and tail olive-brown, sometimes inclining o rufous; lower back dark *chestnut ;* fore-neck naked; chest and breast olive-brown, each feather *margined all round* with whitish-grey; rest of under-parts rufous-brown. Total length, 22 inches; wing, 9·5; tail, 10; tarsus, 2·4; middle toe and claw, 2·3.   As already noted, this species is peculiar in having the chin and throat more or less covered with feathers, in which respect it is approached by some examples of the nearly allied *P. sclateri* and *P. argyrotis*.

Range.—South America; Venezuela, United States of Colombia, and Ecuador.

III. SCLATER'S PENELOPE.   PENELOPE SCLATERI.

Penelope sclateri, G. R. Gray, P. Z. S. 1860, p. 270; Ogilvie-Grant, Cat. B. Brit. Mus. xxii. p. 493 (1893).

Adult.—Differs chiefly from *P. montagnii* in having the feathers of the chest and breast margined with whitish *on the sides only;* the eyebrow-stripes and feathers on the cheeks whiter and more marked; and the rump browner.   In most examples the chin and upper part of the throat are almost naked, but in some these parts are partially feathered.   Total length, 22 inches; wing, 9·5–10·2*; tail, 9·4; tarsus, 2·4; middle toe and claw, 2·3.

Range.—South America; Peru and Bolivia.

Habits.—Dr. J. Stolzmann found this species particularly common in Northern Peru and to the west of the Marañon.   It is rarely met with above 7,500 feet, but at Tamiapampa it was very common at 9,000, and appeared to be found right up to the limits of forest growth.   It is more noisy than the *Aburria* (*Aburria aburri*), and, when flying, gives vent to a sound like "Klou-klou-klou," much like that made by Turkeys, and it is probably from this cause that it derives it local name of "Calaluja" in the provinces of Chota and Jaen.   In the department of Cajamarca it is said to nest at the same time as the *Aburria*, that is in the months of December and January.   As a rule it is not a very shy bird, and easily obtained.

---

* No examples in which the sex has been ascertained have been examined.   Possibly the smaller birds, with a wing measuring 9·5 inches, are females.

IV. GREEY'S PENELOPES. PENELOPE JACUPEBA.

Penelope jacupeba, Spix, Av. Bras. ii. p. 54, pl. lxxi. (1823);
Ogilvie-Grant, Cat. B. Brit. Mus. xxii. p. 494 (1893).
Penelope greeyi, G. R. Gray, P. Z. S. 1866, p. 206, pl. xxii.

Adult Male and Female.—General colour of the upper parts (*including the lower back*) and the chest *dark glossy olive-green;* belly finely mottled with *rufous and black;* primary quills *dark*, like the secondaries; feathers of the forehead, eyebrow-stripes, and cheeks *edged with greyish-white;* those of the hind-neck and mantle similarly marked, but with the pale margins less distinct; those of the fore-neck and chest margined on the sides with white.

Male: Total length, 29 inches; wing, 11·5; tail, 12·1; tarsus, 2·3; middle toe and claw, 2·3.

Female: Rather smaller; wing, 10 inches.

Range.—South America; Para, British Guiana, Rio Negro, and the United States of Colombia.

V. THE GUIANA PENELOPE. PENELOPE MARAIL.

? *Penelope jacupema*, Merr. Beytr. Vog. pt. ii. p. 42, pl. xi. (1786); id. Av. Icones et Deser. ii. p. 39 (1786).
Penelope marail, Gmel. S. N. i. pt. ii. p. 734 (1788); Ogilvie-Grant, Cat. B. Brit. Mus. xxii. p. 495 (1893).
Salpiza marail, Wagler, Isis, 1832, p. 1226.
Penelope purpurascens, J. E. Gray (nec Wagler), Knowsl. Menag. ii. pl. xi. (1846).

Adult Male and Female.—General colour above (*including the lower back*) *very dark bluish-green*, much darker than in any other species of the genus; belly uniform *dark brown;* terminal part of the outer primary quills *pale brown;* eyebrow-stripes and cheeks *dark grey;* feathers of the hind-neck and mantle indistinctly margined on the sides with grey, those of the fore-neck and chest with white. Total length, 32·5 inches; wing, 13·5; tail, 13·7; tarsus, 3·1; middle toe and claw, 3.

Range.—South America ; British Guiana and Cayenne.

VI. ORTON'S PENELOPE.   PENELOPE ORTONI.

Penelope ortoni, Salvin, Ibis, 1874, p. 325 ; Ogilvie-Grant, Cat.
B. Brit. Mus. xxii. p. 496 (1893).

Adult.—Brownish-bronze, with a slight wash of green ; the
feathers of the breast *only* margined with white on the sides.
Total length, 32 inches; wing, 11 ; tail, 9·4 ; tarsus, 2·2 ;
middle toe and claw, 2·1.

I have had no opportunity of examining an example of this
species.

Range.—South America ; Western Ecuador.

The type specimen was obtained by Professor Orton near a
place called Mindo on the western slope of the volcano of
Pichincha, Ecuador.

VII. THE PURPLE PENELOPE.   PENELOPE PURPURASCENS.

Penelope purpurascens, Wagler, Isis, 1830, p. 1110; Meyer,
Vogel-Skel. pt. xviii. pl. clxxviii. (1892); Ogilvie-Grant,
Cat. B. Brit. Mus. xxii. p. 496 (1893).
Salpiza purpurascens, Wagler, Isis, 1832, p. 1226.

Adult Male and Female.—General colour above dull brownish-
olive, glossed with bronze-green, and shading into *purplish* on
the secondary quills and upper tail-coverts ; *top of the head and
cheeks brown;* lower back and rump brown, tinged with bronze ;
under-parts brown, the feathers of the neck, mantle, and
breast edged with white on the sides. Total length, 34·5
inches; wing, 15 ; tail, 14·6 ; tarsus, 3·4; middle toe and
claw, 3·2.

Range.—Central America ; Mexico, Yucatan, Guatemala, and
Honduras.

Mr. G. F. Gaumer says :—" The '*Cojolito*' (in Maya, '*Kosh*')
is abundant only in certain localities. I know of but one
forest in Yucatan (Yak-Jonat) where this bird is found, but in.

this forest I think I have seen 800 or more. It is very shy, lives mostly upon the trees, where it feeds upon fruit and flowers, and also, in times of scarcity of fruit, upon leaves and buds. On discovering a tree laden with its favourite food, it utters a loud yell, which is a signal for all the 'cojolitos' in the forest. In a moment, from every part of the forest come the yells of dozens of other individuals ; and soon the tree is covered with these birds, and in a few minutes it is stripped of its fruit, and the 'cojolitos' fly away to return no more. It has been my fortune twice to be beneath the tree when these birds were feeding. The first time I counted eighty-four birds in one hour and a quarter. The second time fifty-one birds were in the tree, when I shot and brought down eight. The flesh is eaten, though it is much darker and more solid than that of the Kambool."

VIII. BRIDGES' PENELOPE. PENELOPE OBSCURA.

Penelope obscura, Illiger; Temm. Pig. et Gall. iii. pp. 68, 693 (1815) ; Ogilvie-Grant, Cat. B. Brit. Mus. xxii. p. 497 (1893).

Penelope nigricapilla, p. 269, *P. bridgesi,* p. 270, G R. Gray, P. Z. S. 1860.

Adult Male and Female.—General colour of the upper-parts and chest olive-brown, glossed with green and washed with copper on the shoulder-feathers, rump, and upper tail-coverts. From the nearly allied *P. purpurascens* it may be distinguished by having the feathers of the top of the head *margined with grey,* the belly indistinctly mottled with rufous, as well as by its smaller size. From *P. jacupeba,* to which it is also nearly related, it is distinguished by the *uniform brown cheeks* and less marked eyebrow-stripes, which are not continued behind the ear-coverts. Total length, 29·5 inches; wing, 11·3 ; tarsus, 3·3 ; middle toe and claw, 3·2.

Range.—South America ; Uruguay, Southern Brazil (Rio to São Paulo), Paraguay, North Argentina, and Bolivia.

I 2

Q

Habits.—According to Temminck this bird is by no means rare in Paraguay, where it is known under the name of "Yacuhu." It is most commonly met with in the forest regions in the neighbourhood of rivers and lakes. Its cry consists of the syllable *yac*, repeated several times and given forth in a high key, and it occasionally calls "*yacu*," from whence it derives its name.

In Lower Uruguay, Mr. W. B. Barrows found this species, the "Pavo del Monte," limited to the borders and islands of the river, where in heavy growth of timber it was not uncommon, though rarely seen. It has a very harsh, cackling cry. The flesh is much esteemed, and the bird is easily domesticated.

Nest.—Said to be a bulky structure placed in trees.

Eggs.—Said to be eight in number; white, and laid in the month of October.

IX. THE CRESTED PENELOPE. PENELOPE CRISTATA.

Meleagris cristata, Linn. S. N. i. p. 269 (1766).
Penelope cristata, Gmel. S. N. i. pt. ii. p. 733 (1788); Ogilvie-
    Grant, Cat. B. Brit. Mus. xxii. p. 498 (1893).
Salpiza cristata, Wagler, Isis, 1832, p. 1226.
Penelope brasiliensis, Bonap. C. R. xlii. p. 877 (1856).

Adult Male and Female.—General colour above olive, glossed with coppery-green; top of the head and crest *uniform dark brown*; white margins to the feathers of the mantle almost, if not entirely, *absent;* lower back and rump dull chestnut, with a slight greenish gloss; breast dull olive, each feather margined with white on the sides; belly *chestnut*.

Male: Total length, 35 inches; wing, 14·5; tail, 14·5; tarsus, 3·7; middle toe and claw, 3·4.

Female: Rather smaller; wing, 13·5 inches.

Range.—Central America; Southern Nicaragua, Costa Rica,

PLATE XXXVIII

THE CHESTNUT-BELLIED PENELOPE.

and Panama. South America; United States of Colombia and Ecuador.

Habits.—In a note by Dr. Von Frantzius we find the following :—" This beautiful 'Wood-Peacock' prefers the thickest parts of the forest, perching in large numbers on the trees, though at no very great height. It is eagerly sought after on account of its delicious flesh, and the more so because it is easy to shoot. It is often kept in a domestic state, as it will stay in the neighbourhood of habitations without trying to escape. In Costa Rica it is called ' Pava.' "

According to Mr. C. W. Richmond, these birds are common in the forests on the Escondido, where they are usually found in the loftiest trees. They are much hunted by the natives, who call them " Qualms." A hunter is guided almost entirely by the low, prolonged cry of the birds, uttered at times while feeding, as without this clue it is almost impossible to find them.

X. THE BOLIVIAN PENELOPE. PENELOPE BOLIVIANA.

? *Penelope jacucaca*, Spix, Av. Bras. ii. p. 52, pl. 68 (1823).
Penelope boliviana, Bonap. C. R. xlii. p. 877 (1856); Ogilvie-Grant, Cat. B. Brit. Mus. xxii. p. 499 (1893).

Adult Male and Female.—Nearly allied to *P. cristata*, but easily distinguished by having the feathers of the crest and mantle *margined with whitish-grey*, by the general colour of the *breast* and belly being *chestnut*, as well as by the smaller size.

Male : Total length, 27 inches ; wing, 12 ; tail, 13 ; tarsus, 3˙3 ; middle toe and claw, 2˙9.

Female : Smaller ; wing, 10˙5 inches.

Range.—South America. Brazil ; Rio Madeira, Rio Solimoës, Lake Manaqueri. Upper Amazons ; Rio Huallaga. Peru and Bolivia.

Habits.—Mr. Jean Stolzmann writes :—"I have only met with this *Penelope* to the east of the Marañon ; it is rather rare at

Huambo (3,700 ft.), equally so at Chirimoto, where 6,000 ft. seems to be the furthest limit of its orographic distribution. It usually keeps in pairs or in little flocks of two or three pairs. Its affinity with *P. sclateri* shows itself not only in the colouring, but also in the voice, in spite of the fact that the note of our bird is very loud and disagreeable, somewhat resembling the bray of an ass, hence its name of '*Gasnadora.*' Another of its ordinary notes, which I have heard several times in the evening at Huambo, is also disagreeable and strange; I at first attributed it to the *Stentor.* In the daytime it generally hides in the thickets and only comes out at sunset. Besides the name mentioned above, it has another, *Puca-cunga* (puca = red, cunga = neck). I have also several times heard a resounding note, like '*ti-tou-ty*,' repeated several times over, which is more pleasant than the others."

XI. THE WHITE-HEADED PENELOPE. PENELOPE PILEATA.

Penelope pileata, Wagler, Isis, 1830, p. 1109; Des Murs,
   Icon. Orn. pl. 23 (1845); J. E. Gray, Knowsl. Menag.
   pl. 9. (1846) [incorrectly coloured]; Ogilvie-Grant, Cat.
   B. Brit. Mus. xxii. p. 500 (1893).
Salpiza pileata, Wagler, Isis, 1832, p. 1226.

Adult Male and Female.—Distinguished from all the species previously described by having *well-marked black eyebrow-stripes* contrasting strongly with the pale sides of the crown. Crown of head *white* with dark shafts and chestnut tips to the longer feathers; neck and under-parts *dark chestnut;* rest of upper-parts rich glossy olive-green; most of the feathers of the mantle, wing-coverts, fore-neck, and breast edged with white on the sides; lower back and rump washed with dark reddish-brown.

Male: Total length, 31 inches: wing, 12·8; tail, 13·6; tarsus, 3·4; middle toe and claw, 3·2.

Female: Smaller; wing, 12·2 inches.

Range.—South America. Brazil; near Para, Rio Madeira, near the mouth of the Rio Negro, and Rio Vautá.

Eggs.—Rather long ovals; shell white, finely pitted all over. Measurements, 2·9 by 2·05 inches.

XII. THE CHESTNUT-BELLIED PENELOPE. PENELOPE OCHROGASTER.

Penelope ochrogaster, Natterer; Pelz. Orn. Bras. pp. 282, 337 (1870); Ogilvie-Grant, Cat. B. Brit. Mus. xxii. p. 501 (1893).

(*Plate XXXVIII.*)

Adult Male.—Nearly allied to *P. pileata*, the *black eyebrow-stripes* being well marked, but the plumage is altogether paler in colour; feathers of the top of the head reddish-brown, edged on the sides with white; back of the neck *dull olive-brown*, with a slight rufous wash, but *scarcely differing in colour from the mantle;* under-parts light chestnut. Total length, 30 inches; wing, 13·3; tail, 14·2; tarsus, 3·5; middle toe and claw, 3·1.

I have had no opportunity of examining the female of this species, but it is said to resemble the male.

Range.—South America; near Cuyaba, Matto Grosso, Brazil.

XIII. THE BROWN PENELOPE. PENELOPE JACUCACA.

Penelope jacucaca, Spix, Av. Bras. ii. p. 53, pl. 69 (1825); Ogilvie-Grant, Cat. B. Brit. Mus. xxii. p. 501 (1893).
Salpiza jacucaca, Wagler, Isis, 1832, p. 1226.
Penelope superciliaris, J. E. Gray (nec Temm.) Knowsl. Menag. ii. pl. viii. (1846).

Adult Male and Female.—Most nearly allied to *P. pileata* and *P. ochrogaster*, which they resemble in having well-marked black eyebrow-stripes, but easily distinguished from both by having the whole plumage of the *upper- and under-parts dark brown*, the former slightly glossed with green; the feathers of the fore-head, wing-coverts, and breast edged on the sides with white.

Total length, 28 inches; wing, 11·8; tail, 11·1; tarsus, 3·2; middle toe and claw, 3·2.

Range.—South America; Bahia, Brazil, and British Guiana.

XIV. THE BAR-TAILED PENELOPE. PENELOPE ARGYROTIS.

Pipile argyrotis, Bonap. C. R. xlii. p. 875 (1856).
Penelope montana, Licht; Bonap. C. R. xlii. p. 877 (1856).
Penelope lichtensteinii, Gray, P. Z. S. 1860, p. 269.
Penelope argyrotis, Ogilvie-Grant, Cat. B. Brit. Mus. xxii. p. 501 (1893).

Adult.—Easily distinguished from all the other species of *Penelope* previously described by the tail-feathers, which have *a distinct terminal band of rufous*. In other respects it most nearly resembles *P. sclateri*, but the pale eyebrow-stripes are more marked, and the feathers of the mantle, wing-coverts, and breast are edged with *pure* white. Total length, 24 inches; wing, 10–10·9; tail, 10·5; tarsus, 2·3; middle toe and claw, 2·3.

Range.—South America; Ecuador, United States of Colombia, and Venezuela.

XV. THE WHITE-WINGED PENELOPE. PENELOPE ALBIPENNIS.

Penelope albipennis, Tacz. P. Z. S. 1877, p. 746; Ogilvie-Grant, Cat. B. Brit. Mus. xxii. p. 502 (1893).

Adult Male.—Appears to be most nearly allied to *P. jacupeba* and *P. ortoni*, but differs from these and all other species of *Penelope* in having the *first eight primary quills white*, with only the base and tip dusky, and the ninth quill whitish towards the middle. Wing, 13; tail, 13; tarsus, 3·4; middle toe and claw, 3·7.

I have never seen an example of this bird, but from what Mr. J. Stolzmann, the original discoverer of the bird, says, it is clearly a well-marked species, and not a partial albino, as one might at first be inclined to believe.

Range.—South America; vicinity of Tumbez, Western Peru.

Habits.—Mr. J. Stolzmann writes : — " This species, the only representative of the Family found on the Peruvian Coast, has now been almost exterminated. I have only seen it at Tumbez, where, thirty years ago, it was still common, and could be found close to the town ; but, thanks to continual persecution, it has retired into the inaccessible mangroves, where I judged there were not more than fifteen pairs left. I gather, however, from what I have been told, that it is still to be found in all the valleys of the larger rivers from Northern Peru to the valley of Chicama. I have no doubt that it also occurs in the valleys of Lambayeque and Naucho (Rio de Saña), but it is everywhere rare and very shy. They tell me that it also inhabits the mangroves at the mouth of the Zurumilla, which forms the frontier between Peru and Ecuador; and I am not sure that it does not occur in the neighbouring districts of the latter republic. The only certain find for this *Penelope* in the neighbourhood of Tumbez is the Isle of Condeza, one of the numerous little islands in the delta of the river. This island is entirely surrounded by an impenetrable wood of rhizophores, whilst the centre is covered with high bushes. This *Penelope* spends the entire day in the inaccessible thickets, only leaving them at sunrise and sunset to search for food amongst the trees (algarrobes). In January and February, which is the season in which we have looked for this bird, its principal food consisted of the black berries of a bush called *lipe*, but it also appears to eat the shoots of the algarrobes.

"It is one of the most difficult birds to procure, but the easiest time to find it is in the early morning and at dusk, when it is feeding ; at other times of the day it is almost impossible to approach it. We were obliged to go to the island during the night; and, as it was then low tide, we had to take a dozen steps up to our knees in mud. On landing we were attacked by clouds of mosquitos, which abound at this season. . . . Suffice it to say that during eight expeditions I only succeeded in obtaining three shots, out of which two birds were wounded and lost, while the third is now in the Warsaw Museum. About the 10th of January, 1877, my companion (M. Jelski) shot a female, which was quietly sitting on a branch, and noticed that another small bird fell at the same moment; this was a chick about two days old, and a second remained on the branch unhurt. Probably the mother was holding them under her wing, for the one that was killed was on the side nearest the shooter. On the same tree there was a thick nest, loosely composed of dry sticks, and placed at a height of about three metres above the ground. My companion brought this living chick back to the house; we were luckily successful in rearing it." A long and interesting account follows of the bringing up of this bird, which showed remarkable intelligence and affection for its human foster-parents.

Nest.—Placed in a tree, and composed of sticks and twigs.

THE BLACK PENELOPES. GENUS PENELOPINA.

Penelopina, Reichenb. Tauben. p. 152 (1862).

Type. *P. nigra* (Fraser).

Sexes *differ from one another in plumage*. Top of the head feathered, but the feathers *do not* form a crest.

The width of the upper mandible is *greater than* the height. Throat and fore-neck *naked, with a median wattle* in the male.

Tail composed of *twelve* feathers, rather long and rounded; the outer pair being about two-thirds of the length of the middle pair.

First primary flight feather *much the shortest :* eighth about equal in length to the tenth ; ninth slightly the longest.

[In the immature male the wing is of quite a different shape, the first primary flight-feather being shorter than the second,

BLACK PENELOPE.

which is about equal to the tenth, and the sixth is slightly the longest.]

Tarsus longer than the middle toe and claw.

Only one species is known.

I. THE BLACK PENELOPE. PENELOPINA NIGRA.

Penelope niger, Fraser, P. Z. S. 1850, p. 246, pl. xxix.

Penelopina nigra, Reichenb. Tauben, p. 152 (1862); Ogilvie-Grant, Cat. B. Brit. Mus. xxii. p. 503 (1893).

(*Plate XXXVII a*)

Adult Male.--The whole plumage black glossed with dark green or bluish-green ; the under-parts, especially the belly, browner and less strongly glossed. Naked space round eye purple ; throat, fore-part of neck, and large wattle red. Bill, legs, and feet red. Total length, 25 inches ; wing 9·3 ; tail, 11 ; tarsus, 2·8–3 ; middle toe and claw, 2·6–2·8.

Adult Female.—May be distinguished by having the feathers of the crown and back of the neck black edged with brown ; the rest of the upper-parts barred with rufous and black ; the chest sandy-brown, indistinctly mottled with black ; the breast and sides with concentric bars of rufous-buff and dark brown ; the belly brownish-grey, with dusky mottling. Colours of naked skin, &c., and measurements as in the male.

Range.—Central America : the highlands of Guatemala.

Mr. Salvin gives the following note on the peculiar sound that this bird makes when on the wing. He says :—" I well remember being startled by a strange sound when shooting in one of the ravines in the Volcan de Agua in Guatemala. Not at first perceiving whence it arose, I walked on, when the noise was again repeated. I then set about discovering the cause, and soon found that it was produced by a male *Penelopina nigra* which, when flying in a downward direction with out-stretched wings, gave forth a kind of crashing, rushing noise which I likened at the time to the falling of a tree."

THE GUANS. GENUS ORTALIS.

Ortalis, Merrem, Av. Icones et Descr. ii. p 40 (1786); Wharton, Ibis, 1879, p. 450.

Type, *O. motmot* (Linn.).

Sexes similar in plumage.   Top of the head feathered.

The width of the upper mandible is *greater than* the height.

A large naked space round the eye ; *a band of thin feathers down the middle of the throat ; no* wattle.

Tail composed of *twelve* feathers, long and rounded, the outer feathers being shorter than the middle pair.

First primary flight-feather *much shorter than* the tenth ; the sixth slightly the longest.

Tarsus about equal in length to the middle toe and claw.

To facilitate identification the species may be divided as follows :—

A. Outer primary flight-feathers brown or bronze-brown.

  a. Extremities of the outer tail-coverts chestnut.

    a¹. Chest uniform in colour.

      a². Outer tail-feathers chestnut almost to the base (species 1 and 2, pp. 235, 236).

      b². Outer tail-feathers with the basal half or more dark, and the remainder chestnut (species 3 to 6, pp. 236-238).

    b¹. Chest feathers margined or spotted with whitish at the extremity.*

      c². Third pair of tail-feathers never widely tipped with chestnut on both webs; no strongly marked eyebrow-stripe extending backwards down the sides of the head (species 7 to 10, pp. 239-241).

      d². Third pair of tail-feathers widely tipped with chestnut on both webs; white eyebrow-stripes strongly marked and continued

* In *O. superciliaris* the spots on the feathers of the chest are rather faint and indistinct.

backwards down the sides of the neck (species 11, p. 244).

b. Extremities of the outer tail-feathers white or buff (species 12 to 14, pp. 244–247).

B. Outer primary flight-feathers chestnut (species 15 to 17, pp. 248, 249).

A. *Outer primary flight-feathers brown, or bronze-brown. Extremities of the outer tail-feathers chestnut. Chest uniform in colour. Outer tail-feathers chestnut almost to the base.*

I. THE GUIANA GUAN.   ORTALIS MOTMOT.

Phasianus motmot, Linn. S. N. i. p. 271 (1766).

Faisan de la Guiane, D'Aubent. Pl. Enl. ii. pl. 32 [No. 146].

Phasianus katraca, Bodd. Tabl. Pl. Enl. pl. 9 (1783).

Phasianus parraka, Gmel. S. N. i. pt. ii. p. 740 (1788).

Ortalida motmot, Wagler, Isis, 1832, p. 1227.

Ortalis motmot, Salvin, Ibis, 1886, p. 175 ; Ogilvie-Grant, Cat. B. Brit. Mus. xxii. p. 505 (1893).

Adult Male.—Top of the head and nape *dark chestnut ;* upper parts olive-brown, with a rufous tinge in freshly-moulted specimens ; sides of head and fore-neck chestnut ; breast brownish or olivaceous-grey ; belly paler. Total length, 23 inches ; wing, 8·3 ; tail, 10·3 ; tarsus, 2·7 ; middle toe and claw, 2·4.

Adult Female.—Probably similar to the male, but I have not examined any examples in which the sex has been properly ascertained.

Range.—Northern South America ; Cayenne, British Guiana, Rio Negro, and Rio Branco.

Habits.—According to Mr. C. B. Brown the native name of this bird is "Nanaquah." It is easily tamed, readily inter-breeds with domestic fowls, and the hybrids are said to be very pugnacious. Its native name is, of course, derived from its cry of "Nannaquoi."

Nest.—Built of sticks and placed in a low tree.

Eggs.—Four in number, speckled, and little inferior in size to those of a fowl. (*C. B. Brown.*)

II. THE PARA GUAN.   ORTALIS ARAUCUAN.

Penelope araucuan, Spix, Av. Bras. ii. p. 56 (1825).
Penelope aracuan, Spix, Av. Bras. ii. pl. 74 (1825).
Ortalida araucuan, Wagler, Isis, 1832, p. 1227.
Ortalis araucuan, Ogilvie-Grant, Cat. B. Brit. Mus. xxii. p. 506
   .(1893).

Adult Male.—Easily distinguished from *O. motmot* by having the top of the head, nape, and feathered parts of the sides of the head *dark brown*, as well as by its much smaller size. Total length, 17 inches; wing, 6·8; tail, 8·2; tarsus, 1·9; middle toe and claw, 1·8.

Adult Female.—Probably similar to the male; but no female specimens in which the sex has been properly ascertained have been examined by me.

Range.—North-east of South America. Province of Maranhao, and the vicinity of Para.

b². Outer tail-feathers with the basal half or more dark, and the remainder chestnut.

III. THE RUFOUS-HEADED GUAN.   ORTALIS RUFICEPS.

Penelope ruficeps, Wagler, Isis, 1830, p. 1111.
Ortalida ruficeps, Wagler, Isis, 1832, p. 1227.
Ortalis ruficeps, Ogilvie-Grant, Cat. B. Brit. Mus. xxii. p. 506
   (1893).

Adult.—Differs from the three following species of this group in having the *top of the head rufous-chestnut;* from *O. motmot* it is at once distinguished by having the outer tail-feathers blackish, glossed with green and tipped with chestnut, as in *O. wagleri*. Total length, 16·25 inches; tail 7·8.

Range.—Brazil.

PLATE XXXIX.

WAGLER'S GUAN.

IV. WAGLER'S GUAN.   ORTALIS WAGLERI.

Ortalida wagleri, G. R. Gray, List Gallinæ Brit. Mus. p. 12 (1867).
Ortalis wagleri, Ogilvie-Grant, Cat. B. Brit. Mus. xxii. p. 507 (1893).

(Plate XXXIX.)

Adult Male and Female.—Top of the head *brownish-black,* shading into dark grey on the nape; eyebrow-stripes, feathers on the sides of the head and down the middle of the throat paler grey; breast and rest of under-parts *chestnut; four* outer pairs of tail-feathers dark bluish-green, widely tipped with dark chestnut. Total length, 25·5 inches; wing, 9·5; tail, 10·5; tarsus, 3; middle toe and claw, 3.

Range.—Western Mexico; Sinaloa and the Territory of Tepic.

V. JARDINE'S GUAN.   ORTALIS RUFICAUDA.

Ortalida ruficauda, Jardine, Ann. Mag. N. H. xx. p. 374 (1847); id. Contr. Orn. 1841, p. 16, pl. [structure].
Ortalida bronzina, G. R. Gray, List Gallinæ Brit. Mus. p. 11 (1867).
Ortalis ruficauda, Ogilvie-Grant, Cat. B. Brit. Mus. xxii. p. 507 (1893).

Adult Male and Female.—Top of the head and nape uniform *dark grey;* chest brownish-olive; breast and *belly whitish buff;* the *five* outer pairs of tail-feathers tipped with chestnut.

Male: Total length, 24 inches; wing, 8·6; tail, 10; tarsus, 2·5; middle toe and claw, 2·4.

Female: Rather smaller; wing, 8·2 inches.

Range.—Northern South America; Venezuela and the island of Tobago.

Eggs.—Long, oval, white, with a finely grained surface. Measurements, 2·45 by 1·7 inches.

VI. THE GREY-HEADED GUAN.   ORTALIS CANICOLLIS.

Penelope canicollis, Wagler, Isis, 1830, p. 1112.
Ortalida canicollis, Wagler, Isis, 1832, p. 1227.
Ortalida guttata, White, P. Z. S. 1882, p. 627.
Ortalis canicollis, Salvin ; Ogilvie-Grant, Cat. B. Brit. Mus.
xxii. p. 508 (1893).

Adult.—Very similar to *O. ruficauda*, but the chest and upper
breast darker and of a more olive-grey colour ; the chestnut
tips to the tail-feathers much wider, and confined to the *two
outer pairs* of tail-feathers. Total length, 20 inches; wing,
8·8 ; tail, 9·9 ; tarsus, 2·35 ; middle toe and claw, 2·35.

Only female examples have been examined, but the male is
no doubt similar in plumage.

Range.—Central South America ; Paraguay, Rio Parana, Rio
Vermejo, Lower Pilcomayo, Salta and Tucuman in the Argen-
tine Republic, Villa Maria on the Upper Paraguay, Brazil.

Mr. J. Graham Kerr says :—"The ʻCharataʼ is exceedingly
abundant in all the thick forests of the Pilcomayo. They
occasionally descend to the ground to feed, but this is comparatively rare ; usually they remain amongst the upper branches
of the trees, feeding on various fruits. This is a sociable bird,
many being usually found near one another. It is also rather
timid ; but this quality is in great part masked by its intense
curiosity. When one enters a forest in which the ʻCharatasʼ
are not accustomed to the sight of man, they examine the intruder curiously, and call their companions with their soft and
cheepy call-note. If one remains perfectly still all the birds
within hearing collect around, and by answering their call-notes
one can bring them down to within a few feet.

"If one walks up towards a tree in which are some ʻCharatasʼ they first utter their soft call-note to draw their companions'
attention, and then, as one comes nearer, they begin to hop
about uneasily, and their voices rise in pitch by regular grada-

tions until they end in shrill screams, amusingly expressive of fear and timidity.

"At times, more especially just about sunrise, the community of Charatas unite together to produce an extraordinary din. They utter loud and very harsh cries, something like the sound of a gigantic rattle, or of the syllables 'chacarata, chacarata,' from which they get their Guaraní name, 'Charata.' All the birds in one part of the forest uniting in this, the effect is almost deafening. Other companies of birds answer, and on a fine morning in the Chaco, just after sunrise, one hears these Charata-choruses resounding in all directions.

"The Charata is a favourite article of food with the Indians, who attract it by imitating the call-note, and shoot it with bow and arrow. Amongst the Tobas it is called "Cochine," in imitation of its call."

b¹. Chest-feathers margined or spotted with whitish at the extremities. Third pair of tail-feathers never widely tipped with chestnut on both webs. No strongly marked eyebrow-stripes extending backwards down the sides of the head.

VII. THE WHITE-BELLIED GUAN. ORTALIS ALBIVENTER.

Penelope albiventris, Wagler, Isis, 1830, p. 1111.
Ortalida albiventris, Wagler, Isis, 1832, p. 1227.
Ortalis albiventris, Ogilvie-Grant, Cat. B. Brit. Mus. xxii. p. 508 (1893).

Adult.—Top of the head and nape brownish-chestnut; feathers on the sides of the face and upper-parts of the neck *pointed and edged with white;* upper-parts olive-brown (with some bronze or purplish gloss in freshly moulted specimens); lower back mostly chestnut; chest and breast brownish, edged with white; *belly white;* three outer pairs of tail-feathers dark olive-green, with the terminal half chestnut. (In some examples the third pair are also tipped with chestnut.) Total length,

19 inches; wing, 7; tail, 8·5; tarsus, 2·2; middle toe and claw 2·2.

I have not examined any examples of this species in which the sex has been satisfactorily ascertained.

Range.—Eastern South America; Provinces of Pernambuco, Bahia, and Minas Geraës, Eastern Brazil.

Habits.—Concerning this bird Prince Maximilian of Neuwied writes:—"I have not found the Aracuary farther south than the Rio Doce, but from thence northward it is often met with on the Mucuri, Alcobacu, in the Sertong of Bahia, Minas Geraës, and in the swamps and carascos of the Campo Geral. It seems to frequent the secluded interior parts of the forest less than the undergrowth of the woodlands, the catingas, carascos, and the thick tangled bushes of the sea coast, where the vegetation is so thickly interwoven that it is scarcely possible to penetrate it. The birds live here, except in the pairing-season, in small flocks, and one frequently hears the loud peculiar cry of the cock, which consists of several separate broken notes. I often found these birds in pairs among the above-mentioned bushes on the sand. When dislodged by my dogs their harsh cry, with other notes, was immediately heard. I also found them on the shores of the River Ilhéos at the commencement of the undergowth. In the month of January I found young birds of this species already quite strong. The Aracuary is excellent eating, and the breast is well-covered with flesh. Its general habits and mode of life are very similar to those of the other species of the group."

Nest.—Built of twigs and placed in a low tree.

Eggs.—Two or three in number, long ovals, white; surface of shell grained. Measurements, 2·3 by 1·5 inches.

VIII. THE SCALY GUAN.   ORTALIS SQUAMATA.

Ortalida squamata, Lesson, Dict. Sci. Nat. lix. p. 195 (1829).
Ortalis squamata, Ogilvie-Grant, Cat. B. Brit. Mus. xxii. p. 509 (1893).

Adult.—Very closely allied to *O. albiventer*, but the top of the head is more olive, and the feathers of the sides of the face and upper parts of the neck are *rounded* and *uniform olive-brown ;* belly *brownish-white.*

Range.—South-eastern South America. Provinces of Rio Grande do Sul and (?) Santa Catharina.

IX. THE WHITE-FRONTED GUAN. ORTALIS CARACCO.

Ortalida caracco, Wagler, Isis, 1832, p. 1227.

Ortalida adspersa, G. R. Gray, List Gallinæ Brit. Mus. p. 13 (1867).

Ortalis caracco, Ogilvie-Grant, Cat. B. Brit. Mus. xxii. p. 509 (1893).

Adult.—Top of the head *dark grey;* front of the head whitish; feathers of the chest and breast rather narrowly margined all round with whitish. Total length, 22 inches; wing, 8·8 ; tail, 10·5 ; tarsus, 2·4 ; middle toe and claw, 2·4.

No specimens have been examined in which the sex has been ascertained.

Range.—North-western South America ; United States of Colombia and Upper Amazonia.

X. SPIX'S WHITE-FRONTED GUAN. ORTALIS GUTTATA.

Penelope guttata, Spix, Av. Bras. ii. p. 55, pl. 73 (1825).

Ortalida guttata, Wagler, Isis, 1832, p. 1227.

Penelope adspersa, Tschudi, Wiegm. Arch. 1843, p. 386.

Ortalis guttata, Ogilvie-Grant, Cat. B. Brit. Mus. xxii. p. 510 (1893).

Adult Male.—Closely allied to *O. caracco,* but only the *forehead* is *grey ;* the whitish margins to the chest-feathers are narrower, and the measurements are somewhat less. Total length, 19 inches; wing, 7·7 ; tail, 9 ; tarsus, 2·3 ; middle toe and claw, 2·3.

No female examples in which the sex has been ascertained

12   R

have been examined. It is somewhat doubtful whether this and the preceding species are really distinct from one another, but more material is required to settle the question.

Range.—North-western and Central South America ; Upper Amazons, United States of Colombia, Ecuador, Peru, Bolivia, Matto Grosso, and the Rio Madeira.

Habits.—According to Mr. J. Stolzmann, this species is common in the valley of Huayabamba, and occurs up to an elevation of 6,ooo feet at Santa Rosa, but from constant persecution it has become much rarer in the more inhabited parts of the country.   He writes :—"One cannot call it a forest bird, for I have never met it in the depths of the forests, and it keeps more to the outskirts, and is met with along river banks and among tangled undergrowth.   It is generally found in small flocks of three or four pairs which, on sighting a human being, utter a weird and hoarse cry, which they repeat several times with out-stretched necks and enquiring gestures.   It is not naturally a shy bird, and it is only in the more inhabited parts of the country that it becomes wild.   Many times at Huambo, when we were busy working, these birds would fly up and perch on the neighbouring trees, sometimes lighting on the roof of the house, and even on the ladder placed before the window. These visits were especially numerous in wet weather, when numbers of these birds, in company with Pigeons, availed themselves of the dry ground under the verandah of the house. The country people call the places that remain dry during the rains, sheltered by overhanging rocks and such-like, " cal-pares," and it is well known to hunters that these spots are a sure find for birds, more especially Penelopes, Ortalids, and Pigeons.   At Troncopola (Huambo) there was a deserted house without walls, and supported by four pillars.   Here, if approached cautiously, we always, especially in the morning, found Guans or Pigeons, attracted by the dry ground and ashes.

"One day, on returning from shooting, I found a flock of

Ortalis in front of my house; one of them, losing its presence of mind, rushed into my room and perched on a plank, but managed to make its escape through a hole in the wall.

"Their cry, which is uttered in the morning and afternoon, may be heard at a considerable distance, each pair in turn uttering an inharmonious duet. When calling they sit perched on the larger branches close to one another, and whilst one, probably the male, repeats the *hou-dou-gou*, his mate adds after the two first syllables *á-ra-cou*, which together make up the word *hou-dou-á-ra-cou*, of which the middle *a* is the highest and most accentuated note. Throughout the greater part of its range in South America it is known by the same name with slight variations; whilst in the district of Ayacucha it is called "manakáraku," and in the Amazon district "uatáraku."

"In addition to its alarm-note during the nesting-season, I can mention three others. The first may be expressed by the syllable 'kyt.' A second, rarely heard, and which is best expressed by the word 'piou,' is an expression of surprise given vent to as the bird flies up. The third is a piercing cry of despair; one day I heard a bird crying in this way during a whole afternoon.

"The following details, which require confirmation, were supplied me by the natives.

"Several females make one nest in common, for twelve to fourteen eggs are to be found in it, and I have only once seen a female with two chicks. The eggs are white. The nest is placed on the ground, and the natives declare that by removing most of the eggs and only leaving one or two the Ortalids can be induced to go on laying, but I think this is extremely doubtful. It is said that the male will cross with domestic poultry. It is certain that March and April are the nesting-season.

"When walking along horizontal branches, these birds place their feet with the toes turning inwards, like other Penelopes and Pigeons. The flight is heavy and short."

Tschudi, who met with this species in Peru, gives the following note regarding its habits :—" This species lives in flocks
in the more thinly-wooded parts of most of the Peruvian
Montañas. The Indians call it ' Haccha-Nualpa ' (Woodcock).
After sunset a number of these birds will collect for the night
on a large tree and give vent to a piercing shrieking cry, which
is fairly expressed by the syllables ' Ven-aca.' Before sunrise
this cry is repeated, and the flock disperses for the day."

d². *Third pair of tail-feathers widely tipped with chestnut on both
webs; white eyebrow-stripes strongly marked and continued
backwards down the sides of the neck.*

XI. THE WHITE-EYEBROWED GUAN.   ORTALIS SUPERCILIARIS.

Ortalida superciliaris, G. R. Gray, List. Gallinæ Brit. Mus.
    p. 10 (1867).
Ortalis superciliaris, Ogilvie-Grant Cat. B. Brit. Mus. xxii.
    p. 511 (1893).

Adult Male.—Easily distinguished from *O. guttata* and the
preceding species by the *wide whitish eyebrow-stripes*, the *four*
outer pairs of tail-feathers tipped with chestnut, the *third pair*
being almost as widely tipped on both webs as the fourth.
Total length, 17 inches ; wing, 6·8 ; tail, 7·2 ; tarsus, 1·8 ;
middle toe and claw, 1·8.

Only the type specimen, a male, is known.

Range.---South America.   The exact locality is not known.

b. *Extremities of the outer pairs of tail-feathers white or buff.*

XII. THE GREY-HEADED GUAN.   ORTALIS POLIOCEPHALA.

Penelope poliocephala, Wagler, Isis, 1830, p. 1112.
Ortalida poliocephala, Wagler, Isis, 1832, p. 1227.
Ortalis poliocephala, Ogilvie-Grant, Cat. B. Brit. Mus. xxii.
    p. 511 (1893).

Adult Male and Female.—Top of the head and neck dark grey ;

general colour of upper-parts and chest, greyish-olive; breast and belly white; *under tail-coverts rufous-buff;* tail-feathers tipped with buff. Total length, 25 inches; wing, 9·6; tail, 10·8; tarsus, 2·8; middle toe and claw, 2·8.

Range.—Mexico; Rio Armeria, Rio Tupila, Real Arriba, Vera Cruz, Oaxaca, and Tehuantepec.

XIII. THE LESSER GREY-HEADED GUAN. ORTALIS VETULA.

Penelope vetula, Wagler, Isis, 1830, p. 1112.
Ortalida vetula, Wagler, Isis, 1832, p. 1227.
Ortalida maccalli, Baird, B. N. Amer. p. 611 (1860).
Ortalida plumbiceps, G. R. Gray, List Gallinæ Brit. Mus. p. 11 (1867).
Ortalida ruficrissa, Sclater and Salvin, P. Z. S. 1870, p. 538.
Ortalis vetula, Boucard, P. Z. S. 1883, p. 460; Ogilvie-Grant, Cat. B. Brit. Mus. xxii. p. 512 (1893).
Ortalis vetula maccali, Coues; Bendire, N. Am. B. p. 119, pl. iii. fig. 16 [Egg] (1892).
Ortalis vetula pallidiventris, Ridgway, Man. N. Am. B. p. 209 (1887).

Adult Male and Female.—Differs from *O. poliocephala* in being much smaller, and in having the head and neck less grey. Total length, 20 inches; wing, 8; tail, 9·6; tarsus, 2·5; middle toe and claw, 2·5.

This species varies somewhat in colour in the different parts of its wide range, especially on the under-parts of the body, but from the very large series of specimens examined it is clear that all are merely climatic varieties of one form.

Range.—Southern Texas, extending through Eastern Mexico and Central America to the United States of Colombia.

Habits.—Assistant-Surgeon James C. Merrill, U.S. Army, in his notes on the "Ornithology of Southern Texas," writes as follows:—"The 'Chachalac,' as the present species is called on the Lower Rio Grande, is one of the most characteristic birds of that region. Rarely seen any distance from woods or dense chaparral, they are abundant in those places, and their hoarse cries are the first thing heard by the traveller on awaking in the morning. During the day, unless rainy or cloudy, the birds are rarely seen or heard, but shortly before sunrise and sunset they mount the topmost branch of a dead tree, and make the woods ring with their discordant notes. Contrary to almost every description of their cry which I have seen, it consists of three syllables, though occasionally a fourth is added. When one bird begins to cry, the nearest bird joins in at the second note, and in this way the fourth syllable is made; but they keep such good time that it is often very difficult to satisfy oneself that this is the fact. I cannot say certainly whether the female utters this cry as well as the male, but there is a well-marked anatomical distinction in the sexes in regard to the development of the trachea. In the male this passes down the outside of the pectoral muscles, beneath the skin, to within about one inch of the end of the sternum; it then doubles on itself and passes up, still on the right side of the keel, to descend within the thorax in the usual manner. This duplicature is wanting in the female. These birds are much hunted for the Brownsville market, though their flesh is not particularly good, and the body very small for the apparent size of the bird. Easily domesticated, they become troublesomely familiar, and decided nuisances when kept about the house."

Mr. J. A. Singley says:—"All the nests I found were in mesquite stubs, where the limbs had been cut off to make brush-fences. These limbs are never cut close to the tree, and being close together form a cavity; leaves and twigs will fall in this and accumulate, and the bird occupies it as a nesting site. I did not find a nest that I could say was built by the bird. When the nest is approached the bird quietly flies off, rarely remaining in sight, and soon calls up its mate."

Mr. George B. Sennett makes the following statement:—
"The chicks are hatched well-coated with down, and they

leave the nest as soon as hatched, the old ones leading them into the thickets, where they are very hard to capture. I had the pleasure, at the ranch, of seeing six hatch under a hen. The little ones looked and acted exactly like chickens, picking up the corn-batter thrown to them, running in and out from under the hen's wings, and jumping upon her back. Four of the six died within the first two weeks, but the others lived and thrived. A few are domesticated every year at almost every ranch, and they become inconveniently familiar, getting about under foot, jumping upon tables, beds, &c."

Mr. G. F. Gaumer, writing from Yucatan about the "Cha-cha-la-ca," says :—"This bird spends most of its time in the trees, where it lives upon the fruit, flowers, and tender leaves. Its neutral green plumage renders it very difficult to spy out the bird. When disturbed it jumps to the ground to ascertain the nature of its danger, gives one or two long leaps, and again mounts upon a limb, from which it quickly flies from one branch to another until it escapes in the distance. In the male the trachea is wonderfully prolonged beneath the skin of the breast and abdomen almost to the anus, whence it returns and enters the chest at the proper place. With the great trumpet-like instrument the bird makes a peculiar noise, which may be heard at a league's distance. The song is harsh and sonorous, and never produced alone; but after each part the female, with a finer shriller voice, repeats it in such rapid succession, that it seems like one bird doing the whole. The usual time of singing is in the morning and evening, but it frequently sings at other hours."

Eggs.—Generally three or four in number, occasionally five, longish ovals; shell, creamy-white, finely pitted. Measurements, 2·3 by 1·6 inches.

XIV. THE WHITE-VENTED GUAN.   ORTALIS LEUCOGASTRA.

Penelope albiventer, Gould (*nec* Wagler), Voy. Sulph. Zool. p. 48, pl. 31 (1844).

Penelope leucogastra, Gould, P. Z. S. 1843, p. 105.
Ortalida leucogastra, G. R. Gray, List Gallinæ Brit. Mus.
   p. 20 (1844).
Ortalis leucogastra, Ogilvie-Grant, Cat. B. Brit. Mus. xxii.
   p. 514 (1893).

Adult Male and Female.—Easily distinguished from the two
preceding species by having the under tail-coverts *white.* Total
length, 20·5 inches ; wing, 8·3 ; tail, 8·5 ; tarsus, 2·3 ; middle
toe and claw, 2·3.

Range.—Central America; Nicaragua, Salvador, and the
Pacific slope of Guatemala.

Habits.—Mr. Salvin says :—"This Guan is very abundant in
the Pacific coast region, where in the neighbourhood of the
more remote and smaller villages, the woods in the early morning resound with its loud continued cries. It is usually seen in
trees, and shows little symptom of alarm on one approaching.
The time of breeding seems to extend over some period, as
young birds and fresh eggs were observed simultaneously in
the month of March. The former appear to run almost
immediately on becoming free from the shell, and, clinging
to the branches of the underwood, are nimble in eluding
capture."

Nest.—Is usually placed in a low bush, and is composed
entirely of small twigs.

Eggs.—Two in number and of a rough texture ; pure creamy-
white in colour. Measurements, 1·5 by 1·25 inch.

   B.—*Outer primary flight-feathers chestnut.*

XV. THE CHESTNUT-WINGED GUAN.   ORTALIS GARRULA.

Phasianus garrulus, Humb. Obs. de Zool. i. p. 4 (1811).
Penelope garrula, Wag'er, Isis, 1830, p. 1111.
Ortalida garrula, Wagler, Isis, 1832, p. 1227.
Ortalis garrula, Ogilvie-Grant, Cat. B. Brit. Mus. xxii. p. 515
   (1893).

Adult Female.—Top of the head and nape dull *chestnut;* rest of the upper-parts greyish-olive ; chest olive-grey, shading into white on the rest of the under-parts ; the *five* outer pairs of tail-feathers *tipped with white or pale buff.* Total length, 22 inches ; wing, 8·8 ; tail, 9·6 ; tarsus 2·8 ; middle toe and claw, 2·7.

The plumage in the male and female is probably similar, but no male example, in which the sex has been ascertained, has been examined.

Range.—Northern South America ; the coast region of the United States of Colombia, and near Caracas, Venezuela.

XVI. THE GREY-HEADED CHESTNUT-WINGED GUAN.
ORTALIS CINEREICEPS.

Ortalida poliocephala, Auct. (*nec* Wagler).

Ortalida cinereiceps, G. R. Gray, List Gallinæ Brit. Mus. p. 12 (1867).

Ortalida frantzii, Cabanis, J. f. O. 1869, p. 211.

Ortalis cinereiceps, Richmond, P. U. S. Nat. Mus. xvi. p. 523 (1893) ; Ogilvie-Grant, Cat. B. Brit. Mus. xxii. p. 515 (1893).

Adult Male and Female.—Similar to *O. garrula,* but the head and nape are *dark grey.* Total length, 22 inches ; wing, 8·5 ; tail, 8·7 ; tarsus, 2·8 ; middle toe and claw, 2·6.

Range.—Central America ; Costa Rica, Veragua, Panama, to the United States of Colombia.

Habits.—Dr. Von Frantzius writes that this bird is "universally called 'Chachalaca,' and is spread over the whole of the high plateau, particularly in open places near the forest."

XVII. THE ECUADOR CHESTNUT-WINGED GUAN.
ORTALIS ERYTHROPTERA.

Ortalida erythroptera, Natterer MS. ; Licht. Nomencl. p. 87 (1854) [descr. nulla] ; Sclater and Salvin, P. Z. S. 1870, p. 540.

Ortalis erythroptera, Ogilvie-Grant, Cat. B. Brit. Mus. xxii. p. 516 (1893).

Adult.—Easily distinguished from the two preceding species by having the *four* outer pairs of tail-feathers *widely tipped with dark chestnut.* From *O. garrula*, which it otherwise resembles, it may be further distinguished by having the top of the head and nape brighter chestnut, and the *under tail-coverts of a paler chestnut colour.* Total length, 24 inches; wing, 9·2; tail, 10·6; tarsus, 3·1; middle toe and claw, 3·1.

Range.—North-western South America; Babahoyo, Guayaquil, and Palmal, Western Ecuador. ? Cumana, Venezuela.

THE PIPING GUANS. GENUS PIPILE.

Pipile, Bonap, C. R. xlii. p. 877 (1856).

Type, *P. cumanensis* (Jacquin).

The width of the upper mandible *greater than* the height.

Sexes similar in plumage. A well-developed crest of pointed feathers. A large patch round the eyes naked. Front of the neck almost naked, with a *median wattle.*

Tail composed of *twelve* feathers, rather long and rounded, the outer pair being distinctly shorter than the middle pair.

Two outer primary flight-feathers with the last third of the inner web *deeply excised.* The first much *shorter than* the second, which is about equal to the tenth; sixth slightly the longest.

Tarsus *shorter than* the middle toe and claw.

I. THE WHITE-HEADED PIPING GUAN. PIPILE CUMANENSIS.

Crax cumanensis, Jacquin, Beytr. p. 25, pl. 10 (1784).
Crax pipile, Jacquin, Beytr. p. 26, pl. 11 (1784).
Penelope leucolophos, Merrem, Av. Icones et Descr. ii. pp. 43, 44. pl. 12 (1786); id. Beytr. Vög. ii. pp. 46, 47, pl. 12 (1786).

Penelope cumanensis and *P. pipile*, Gmel. S. N. i. pt. ii. p. 734 (1788); J. E. Gray, Knowsl. Menag. ii. pl. 10 (1846).

Pipile cumanensis, Bonap.; Ogilvie-Grant, Cat. B. Brit. Mus. xxii. p. 517 (1893).

Pipile nattereri, Reichenb. Tauben, p. 154 (1862).

Penelope jacquinii, G. R. Gray, List Gallinæ Brit. Mus. p. 8 (1867).

Penelope grayi, Pelz. Orn. Bras. p. 284 (1870).

Adult Male and Female.—Top of the head and crest *white*. General colour above black, glossed with *dark green*; the first six or seven *outer secondary coverts white on both webs*, except the tips, which are black glossed with green; some of the feathers of the chest and breast margined with white.

Male: Total length, 30 inches; wing, 13·2; tail, 11·5; tarsus, 2·5; middle toe and claw, 2·6.

Female: Smaller; wing, 12·6 inches.

Range.—South America : British Guiana, Venezuela, Trinidad, United States of Colombia ; Rio Negro, Upper Amazons ; Rio Napo, Ecuador; Eastern Peru, Bolivia; and Matto Grosso in Brazil.

II. THE BLACK-FRONTED PIPING GUAN.   PIPILE JACUTINGA.

Penelope jacutinga, Spix, Av. Bras. ii. p. 53, pl. 70 (1825).

Penelope nigrifrons, Temm. MS. ; Lesson, Traité d'Orn. p. 482 (1831).

Penelope leucoptera, Neuwied. Beitr. Nat. Bras. iv. p. 544 (1832).

Pipile jacutinga, Sclater and Salvin, P. Z. S. 1870, p. 530; Meyer, Vogel-Skel, pt. xiv. pl. cxxxvii (1890); Ogilvie-Grant, Cat. B. Brit. Mus. xxii. p. 518 (1893).

Adult Male and Female.—General colour above brownish-black *glossed with purple;* forehead and eyebrow-stripes *black;* crest white with black shafts; outer webs *only* of the secondary coverts white; chin and upper part of throat covered with black feathers; naked space round the eye much smaller than in *P. cumanensis* and the white margins to the feathers of the breast more strongly marked. Total length, 30 inches; wing, 13; tail, 11·5; tarsus, 2·5; middle toe and claw, 2·7.

Range.—Eastern South America : Bahia, São Paulo, Rio de Janeiro, Rio Parana, Rio Grande do Sul, and Paraguay.

Habits.—Herr Bischoff, writing from Arroio Grande, says that the Jacutinga is a migratory bird, arriving there in May and June in flocks of from four to sixteen individuals. It nests in trees, selecting a part of the stem where three or four branches arise, and depositing its eggs in this natural hollow without any lining or attempt at a nest. . . . He once had the opportunity of observing the nesting habits of the Jacutinga, and both the male and female appeared to take part in the incubation. The young were hatched in November, and could soon not only follow their parents, but fly. In December they depart from Arroio Grande. They cannot be domesticated, for they are most pugnacious and kill poultry.

From Prince Maximilian of Neuwied's excellent work* we learn that the "Jacutinga" is met with only in the more secluded parts of the vast forests, and is generally found singly or in pairs. Its habits are similar to those of the Jacupemba (*Penelope superciliaris*), but its call is short and shrill, and so far as he recollected the trachea is more highly developed. It can be tamed and soon becomes quite domesticated, while, in the interior of the forest, its flesh is a valuable addition to the larder. Unlike the Jacupemba, it is never to be seen near the sea coast, and its food appears to consist of fruits and insects, judging from the remains found inthe crop of specimens examined.

The Indians use the large strong tail-feathers to make wings for their arrows.

* Beitr. Nat. Bras. iv. p. 544 (1832).

Nest.—A nest found in the month of February was placed in a tree, and composed of twigs. As will be seen above, it would appear that in some cases no nest is made.

Eggs.—Two or three in number ; white, and as large as those of a Turkey.

III. THE AMAZONIAN PIPING GUAN.   PIPILE CUJUBI.

Yacon Turkey, Latham, Gen. Syn. ii. pt. ii. p. 681, pl. lxi (1783).
Penelope cujubi, Natterer MS. ; Pelz, SB. Ak. Wien, xxxi.
    p. 328 (1858) ; id. Orn. Bras. p. 284 (1870).
Pipile cujubi, Reichenb. Tauben, p. 153 (1862) ; Ogilvie-
    Grant, Cat. B. Brit. Mus. xxii. p. 519 (1893).

Adult Male and Female.—Like *P. jacutinga*, the general colour of the upper-parts is brownish-black, glossed with purple, but the feathers of the crest are dark brown, edged with white; the chin and throat almost naked ; and the outer webs of the secondary coverts *dark brown, margined* with white.

Male : Total length, 30 inches ; wing, 13·5 ; tail, 11·5; tarsus, 2·55 ; middle toe and claw, 2·7

Female : Smaller ; wing, 12·7 inches.

Range.—North-eastern South America ; Lower Amazons and Para.

THE WATTLED GUANS.   GENUS ABURRIA.

Aburria, Reichenb. Nat. Syst. Vögel, p. xxvi. (1852).

Type, *A. aburri* (Less.).

Sexes *similar* to one another in plumage.

The width of the upper mandible *greater than* the height. Fore part of neck *mostly feathered*, with a *long vermiform wattle*. Only a small naked space below the eye. Tail composed of *twelve* feathers.

First three primary quills deeply excised at the extremity, the

fourth less so. First primary much shorter than the second;
fourth equal to the tenth, and the sixth slightly the longest.
Tarsus longer than the middle toe and claw.

I. THE BLACK WATTLED GUAN. ABURRIA ABURRI.

Penelope aburri, Lesson, Dict. Sci. Nat. lix. p. 191 (1829).
Aburria carunculata, Reichenb. Syst. Av. p. xxvi.(1852); Tacza-
    nowski, Orn. Pérou. iii. p. 277 (1886).
Aburria aburri, Ogilvie-Grant, Cat. B. Brit. Mus. xxii. p. 520
    (1893).

Adult Male and Female.—Whole plumage black, glossed with
dark green. Total length, 29 inches; wing, 14–14·5 ; tail, 12 ;
tarsus, 2·7 ; middle toe and claw, 2·9.

Range.—Western South America ; the interior of the United
States of Colombia, Ecuador, and Northern Peru.

Mr. J. Stolzmann found the Black Wattled Guan very common at Tambillo, but it appeared to become scarcer to the east
of the Marañon, and much wilder in the Amazon District. At
Tambillo it is most common in the valley, and becomes rarer
as the higher altitudes are reached, where its place is taken by
Penelope sclateri, though it was obtained at an elevation of
7,000 feet. At Huambo it was decidedly scarce.

"Its cry," he says, "is the most curious that I have heard. It
begins with a note repeated several times, resembling the sound
of a trumpet, ascending in semi-tones that can only be heard
when one is close at hand. Then it gives vent to a very loud
guttural cry, which begins in a low key, and, gradually ascending
in quick time to a high note, again descends. This cry reminds
one of that of the Crane. As a rule it flies without making any
noise, so much so, indeed, that it frequently escapes unob-
served. Its cry is most often heard in the nesting-season, that
is, between the months of September and February ; and when
calling it remains stationary and in a crouching position. The

female defends her young with great spirit; for one day, while traversing the forest, a female of this species attempted to stop me, passing so close that I could almost have caught her in my hand; then with loud cries, having attacked my dog, almost striking him with her wings, she took up a position on a neighbouring tree. It was not till I had shot her that I perceived that she had been defending her young, for on searching about I soon discovered her empty nest, and next day found one of her chicks in the same neighbourhood.

"During moonlight nights one often hears their strange cries. At Cocochó they are hunted on such nights, for at other times they are very wild, and their flesh is much esteemed. They feed on the ground, and, when the fruits of the "Nectardes" are ripe, feed principally on them.

"One generally finds these birds singly or in pairs, and I have only once met with a flock of a dozen birds. The inhabitants usually call this bird *Pava negra*, a name applied to *Pipile* at Moyobamba, where the Aburri receives the name 'Uante.' When only winged, these birds run so quickly that they are frequently lost to the sportsman."

Nest.—Placed on the top of a low tree, and principally composed of sticks and leaves.

Eggs.—Usually two in number. Broad ovals; shell pure white, grained. Measurements, 2·65 by 2 inches.

THE SICKLE-WINGED GUANS. GENUS CHAMÆPETES.

Chamæpetes, Wagler, Isis, 1832, p. 1227.

Type, *C. goudoti* (Less.).

Sexes *similar* in plumage.

The width of the upper mandible *greater than* the height. The fore-part of the neck, as well as the chin and throat, *entirely feathered*. *No* wattle. *A naked space* round the eye.

Tail composed of twelve feathers, long and rounded, the outer pair being a good deal shorter than the middle pair.

The two outer primary flight-feathers deeply excised at the extremity, and the third less so. First primary much shorter than the second; fourth about equal to the tenth; sixth or seventh slightly the longest.

Tarsus slightly shorter than the middle toe and claw.

I. THE RUFOUS-BREASTED SICKLE-WINGED GUAN.
CHAMÆPETES GOUDOTI.

Ortalida goudotii, Lesson, Man. d'Orn. ii. p. 217 (1828).
Chamæpetes goudotii, Wagler, Isis, 1832, p. 1227; Ogilvie-Grant, Cat. B. Brit. Mus. xxii. p. 521 (1893).
Penelope rufiventris, Tschudi, Wiegm. Arch. 1843, p. 386.
Chamæpetes tschudii, Taczanowski, Orn. Pérou, iii. p. 275 (1886).

Adult Male and Female.—General colour above *brownish*, glossed with bronze-green; the lower chest *cinnamon* shading into *rufous* on the breast, and *chestnut* on the flanks. Total length, 24 inches; wing, 10·2; tail, 10; tarsus, 2·6; middle toe and claw, 2·7.

Range.—Western South America. Interior of the United States of Colombia, Ecuador, and Peru.

Habits.—Mr. J. Stolzmann writes:—" I only found this Guan east of the Marañon; it is not met with in the provinces of Chota and Jaen. At Tamia-pampa it occurs up to an elevation of 9,000 feet, and I was told that it is also to be found at Huambo, but is rare in that locality. It is most common at an elevation of about 6,000 feet. In the Amazon district it is known under the name of ' *Pischa*.' "

Eggs.—Perfectly oval; shell pure white, rather glossy and very finely pitted. Measurements, 2·75 by 1·95 inches.

II. THE BLACK-BREASTED SICKLE-WINGED GUAN.
CHAMÆPETES UNICOLOR.

Chamæpetes unicolor, Salvin, P. Z. S. 1867, pp. 159, 160; Ogilvie-Grant, Cat. B. Brit. Mus. xxii. p. 522 (1893).

Adult Male and Female.—General colour above *black*, glossed with dark green; under-parts similar, but browner on the belly, and generally indistinctly mottled with pale rufous-buff. Total length, 25 inches; wing, 11 to 11·7; tail, 10·5; tarsus, 2·7; middle toe and claw, 2·85.

Range.—Central America; Costa Rica and Veragua.

Habits.—The following note is taken from Dr. v. Frantzius' paper ("Journal fur Ornithologie," 1869, p. 372):—"This species, locally known as the *Gallina volcanica* has, up to the present time, only been obtained on the slopes of Irazu, where it is very common, especially near Rancho Redondo and La Palma. It is often brought into the towns to be sold, and is most frequently shot at the end of the rainy season, as it then leaves the thick forests on the hills and visits the lower-lying and less densely-wooded parts."

THE HOATZINS. ORDER OPISTHOCOMI.

Although only a single species comprises this Order, the structure of the skeleton presents so many important peculiarities that the Hoatzin can only be placed in an isolated position. The modifications of the alimentary tract are almost as remarkable as the skeletal characters.

With regard to the latter, Professor Huxley says that it "resembles the ordinary gallinaceous birds and pigeons more than it does any others, and when it diverges from them it is either *sui generis*, or approaches the *Musophagidæ*." The latter group, known as the Touracos, and the Cuckoos (*Cuculidæ*), are, according to the late Professor Garrod, nearly allied.

12   S

Among other peculiar skeletal characters may be mentioned the schizognathous palate, the absence of basipterygoid processes (*cf.* vol. i. p. 1, fig. 1), the shape of the dorsal vertebræ and the sternum. The shape of the latter is unique, the lateral edges being nearly parallel for about two-thirds of their length, then diverging so that the sternum is wider posteriorly than anteriorly. There are two small notches on either side of the posterior margin, the outer being reduced to a foramen. The kel of the sternum is very small, and cut away in front with a flattened out and broadened surface at the posterior termination. On this flattened surface the greater part of the weight of the body is supported when the bird is at rest. The bones of the shoulder girdle—the coracoids, clavicles, and furcula—are completely joined (anchylosed) to one another and to the sternum.

The crop is enormous, and occupies the upper part of the chest, distorting the furcula and sternum ; it is placed in a deep cavity in the upper half of the pectoral muscles.

The hind toe, or hallux, is very long.

The first secondary quill is not much shorter than the second.

The oil-gland is tufted.

The young are hatched naked, the thumb and first finger being provided with claws which enable them to climb and grasp the branches soon after they are hatched ; the bill, as well as the legs and wings, being used for a similar purpose. They are able to swim and dive with facility, when compelled to do so.

The eggs are double-spotted and remarkably Rail-like.

FAMILY OPISTHOCOMIDÆ.

THE HOATZINS. GENUS OPISTHOCOMUS.

Opisthocomus, Illiger, Prodr. Syst. Mamm. et Av. p. 239 (1811).

Type, *O. hoazin* (Müll.).

PLATE XL.

Wyman & Sons Limited

HOATZIN (Young.)

Sexes similar in plumage.

An elongate crest of rather stiff-shafted feathers sides of the head mostly naked.

Tail composed of *ten* feathers, elongate and rounded; the middle pair considerably longer than the outer pair.

First primary flight-feather much shorter than the second, which is shorter than the tenth; sixth to eighth longest.

Fifth secondary quill present.

Tarsus with reticulated scales, shorter than the middle toe and claw.

I. THE HOATZIN. OPISTHOCOMUS HOAZIN.

Phasianus hoazin, Müll. S. N. Suppl. p. 125 (1776).

Crested Pheasant, Latham, Gen. Syn. ii. pt. ii. p. 720, pl. lxiv. (1783).

Phasianus cristatus, Gmel. S. N. i. pt. ii. p. 741 (1788).

Opisthocomus hoatzin, Steph. in Shaw's Gen. Zool. xi. p. 193 (1819).

Opisthocomus cristatus, L'Herminier, C. R. v. p. 433 (1837); G. R. Gray, Gen. B. ii. p. 396, pl. xcvii. (1845); Huxley, P. Z. S. 1867, pp. 435, 460, 1868, p. 304 [skeleton], fig. 4, pl. xcviii. (1845); Cabanis, J. f. O. 1870, p. 318, pl. i. fig. 3 [egg]; Perrin, Trans. Z. S. ix. p. 353, pls. 63–66 (1875); Garrod, P. Z. S. 1879, p. 109 [anatomy]; Quelch, Ibis. 1890, p. 327 [habits]; Parker, Trans. Zool. Soc. xiii. pp. 43–86, pls. vii.-x. (1891); Gadow, P. R. Irish. Ac. (3) ii. p. 147, pls. vii. viii. (1892).

Opisthocomus hoazin, Ogilvie-Grant, Cat. B. Brit. Mus. xxii. p. 524 (1893).

(*Plate XL.*)

Adult Male and Female.—General colour above dark brown glossed with olive; fore part of the head reddish-brown; long crest-feathers mostly dark brown; feathers on the back of the neck and mantle similar, but the former with buff and the latter with white shaft-stripes; the shoulder-feathers and wing-coverts either margined or tipped with white; quills chestnut-brown;

S 2

under-parts pale buff, shading into chestnut on the sides and
belly; tail-feathers widely tipped with pale buff. Total length,
23 inches; wing, 12·4–12·6; tail, 11·4–12·4; tarsus, 1·9–2·2;
middle toe and claw, 1·9.

Range.—South America; Surinam to the United States of
Colombia, and southwards to Bolivia.

Habits.—Mr. J. J. Quelch says:—"The Hoatzin is known
in British Guiana by the various names of 'Anna,' 'Hanna,'
'Canjé, or Stinking Pheasant,' and 'Govenor Battenberg's
Turkeys'; but in the districts where it is found, the name of
'Hanna' is the one most commonly used. . . .

"In the early morning or in the late afternoon they will be
seen sitting in numbers on the plants, while towards the middle
of the day, as the fierce heat of the sun increases, they betake
themselves to shelter, either in the denser recesses of the
growths, or among the individual trees of denser foliage, or
among the tangled masses of creeping and climbing vines,
which frequently spread over considerable areas of their food-
plants along the very edge of the water. . . . Late in
the evening, after feeding, they will be seen settling themselves
down in suitable places for the night.

"The cry of the Hoatzins is easily heard when they are disturbed, and it is one of which it is not easy to give an exact
idea. It recalls slightly the shrill screech of the Guinea-bird
(*Numida*), but it is made up of disjointed utterances, like the
notes 'heigh' or 'sheigh' (*ei* as in sleigh), pronounced with a
peculiarly sharp and shrill nasal intonation, so as to be quite
a 'hiss.' While they are treading, the noise made is
considerable, the cry being more continuous and shriek-
like. . . .

"The nesting-time of the birds certainly extends from
December to July, and I think it very likely that it is continuous throughout the year. . .

"The nests, which are made solely of a slightly concave

mass of dried twigs and sticks taken from the plants on which
they are built, and loosely laid on top and across each other,
are placed in conspicuous positions high up over the water or
soft mud, on the top of or amongst the bushy growth, where
they are fully exposed to the direct sunshine. Almost invariably the plants thus built on were the close-growing 'Bun-
doorie pimpler,' though in a few cases I have seen them on
the Courida, and on a Pimpler (or prickly) palm (*Bactris
major*). . . .

"From the binding nature of the spiny twigs, the nests last
for a considerable time, and these are certainly made use of
again, possibly after more or less repair. The same nest has
been found in use after an interval of seven months.

"Two or three eggs are laid in a nest, both numbers being
about equally common in my experience, and in one special
case six eggs were taken from a nest on which one bird had
been sitting, but whether they had been laid by one or two
birds there was nothing to show. The eggs are easily seen
from beneath the nest, owing to its loose structure. . . .
Even while the birds are sitting on them the eggs must be kept
fairly cool from below; and this evidently gives the explana-
tion why a number of freshly-laid eggs that were placed to be
hatched out by a common fowl exploded one after another,
much to the alarm of the foster-parent, who, however, stuck to
the nest with the remainder after each occurrence.

"Soon after the hatching of the eggs, the nestlings begin to
crawl about by means of their wings and legs, the well-
developed claws on the pollex and index being constantly in
use for holding and hooking on to the surrounding objects.
If they are drawn from the nest by means of their legs they
hold on firmly to the twigs both by their bill and wings.
When the parent bird is driven from the nest owing to the
close approach of a boat, the young birds, unless they are
only quite recently hatched, crawl out of the nests on all fours,
and rapidly try to hide in the thicker bush behind.

"One curious feature noticed with a nestling, which had been upset into the river, was its power of rapid swimming and diving, when pursued. As soon as the hand was placed close to it, it rapidly dived into the dark water, in which it was impossible to see it, and would rise at distances of more than a yard away. Owing to this power the little creature managed to evade all my attempts to seize it, taking a refuge eventually far under the bushy growth, where it was impossible to pursue it. The prolonged immersion which a nestling will thus instinctively and voluntarily undergo, or which an adult bird will bear in an attempt to drown it, seems to me quite remarkable."

Nest.—Described above.

Eggs.—Two appears to be the general number laid, or, in some cases, three. As many as six are recorded in a single instance, probably the result of two birds laying in the same nest. Oval in shape, and remarkably like those of the Common Corn-crake. Ground colour pale buff or stone colour, marked with pale cloudy violet undermarkings and reddish-brown surface spots. Average measurements. 1·8 by 1·3 inch.

THREE-TOED QUAILS AND PLAIN-WANDERERS.
ORDER HEMIPODII.

The small Quail-like birds comprising this Order occupy a somewhat isolated position, and are not really very closely allied to the Gallinaceous birds, though, perhaps, nearer to them than to any other group. Some of their characteristics show a marked affinity to the Rails, and on the whole the most natural position for the Hemipodii appears to be one between these two great Orders of birds.

Among the important skeletal characters may be noted the following :—

In the skull the maxillo-palatine bones are not coalesced with one another, nor with the vomer (see vol. i. p. 1, fig. 1) ; the nasals are schizorhinal (fig. 2). The vertebræ are peculiar in shape, and of the form known as heterocœlous. There is a deep notch on each side of the posterior margin of the sternum, extending for about two-thirds of its entire length, and the well-developed episternal process is incompletely perforated to receive the bases of the coracoid bones, which are only separated by a thin bony septum.

Bill like that of the *Gallinæ*, but often not so strongly developed.

Feet generally with three toes only, the hind toe, or hallux, being absent, except in *Pedionomus*, which possess a rudimentary hind toe.

Oil-gland tufted.

Tail very short, composed of soft feathers scarcely to be distinguished from the longer upper tail-coverts.

The first secondary quill not much shorter than the second, and the fifth present.

The young hatched covered with down, and able to run soon after they are hatched.

Eggs double-spotted, and three to five in number.

One of the great peculiarities of this group is the fact that the female is always larger and generally more handsomely marked than the male, and the latter in the majority of species, probably in all, incubates the eggs and tends the young.

FAMILY TURNICIDÆ.

THREE-TOED OR BUSTARD-QUAILS. GENUS TURNIX.

Turnix, Bonnat. Tabl. Encycl. Méth. i. pp. lxxxii. 5 (1890).

Type, *T. sylvatica* (Desf.).

Hind toe, or hallux, absent.

In all the birds of this genus there is a general tendency to uniformity of coloration in the plumage of the upper surface of very old examples; the bars, spots, and markings gradually disappearing with age.

To facilitate identification the twenty-one species and two sub-species comprising this genus may be divided as follows:—

I. Leg (metatarsus) longer than the middle toe and claw.

   A. Entire breast transversely barred with black; belly immaculate; sexes different in plumage.

      a. Chin and throat black or barred with black (females).

      b. Chin and throat white, the feathers on the sides narrowly edged with black (males) (species 1 to 4, pp. 265–270).

   B. Middle of breast not transversely barred with black; throat never black. (Plumage of sexes similar in one group, but slightly different in the other section.)

      c. Middle feathers of the tail lengthened, pointed a.d edged with white or buff.

         a^1. Feathers of the mantle and back edged with white or buff, giving the back a scaly appearance. Plumage of sexes practically similar.

         a^2. Middle of breast and belly, immaculate; sides spotted or barred with black (species 5–7, pp. 270–275).

         b^2. Middle of breast and belly with round black spots on most of the feathers (species 8, p. 275).

         b^1. Feathers of the mantle and back practically uniform (species 9, p. 276).

      d. Middle tail-feathers not lengthened and pointed, nor edged with white or buff; feathers of the back without any scaly appearance. Sexes somewhat different in plumage.

c^1. Shoulder - feathers not edged with golden-buff. (Females with a well-defined rufous collar; males without) (species 10 to 12, pp. 277-280).

d^1. Shoulder-feathers edged with golden-buff (species 13 to 15, pp. 281-282).

C. Neck and breast uniform bright rufous; upper tail-coverts very long, entirely covering the true tail (species 16, p. 283).

II. Leg (metatarsus) equal to or shorter than the middle toe and claw. Bill very stout in some species (species 17 to 22, pp. 284-289).

I. *Tarsus longer than the middle toe and claw.*

A. *Entire breast transversely barred with black; belly immaculate; sexes different in plumage; middle of chin and throat black, or barred with black in the females, white in the males.*

1. THE BUSTARD-QUAIL. TURNIX TAIGOOR.

Hemipodius taigoor, Sykes, P. Z. S. 1832, p. 155; id. Trans. Z. S. ii. p. 23, pl. iv. (1841).

Hemipodius plumbipes, Hodgson, Bengal Sport. Mag. May, 1837, p. 346.

Turnix ocellatus, Jerdon (*nec* Scop.), B. Ind. ii. p. 597 (1864).

Turnix taigoor, Jerdon, B. India, ii. p. 595 (1864); Hume & Marshall, Game Birds of Ind. ii. p. 169, pl. (1879); Oates, ed. Hume's Nests & Eggs, Ind. B. iii. p. 367 (1890); Ogilvie-Grant, Cat. B. Brit. Mus. xxii. p. 531 (1893).

Turnix rostrata, Swinhoe, Ibis. 1865, p. 543.

Areoturnix blakistoni, Swinhoe, P. Z. S. 1871, p. 401.

Turnix plumbipes, Hume & Marshall, Game Birds Ind. ii. p. 177, pl. (1879).

Adult Male.—Chin and throat *white*, narrowly edged with

black on the sides only; chest barred with black and *buff;* belly and thighs rufous; upper-parts rufous or brown; no rufous nuchal collar *contrasting with* the colour of the back. Total length, 5·6 inches; wing, 3·1; tail, 1·2; tarsus, 0·85.

Adult Female.—Chin and throat (and in very old examples the middle of the chest) *black;* no rufous collar round the back of the neck contrasting with the back. General colour above rufous or greyish-brown or any intermediate colour (the colour varying according to the amount of rainfall in the district where the individual occurs, the most rufous forms being found in the more arid localities where the rainfall is small), mottled with black and more or less margined with pale buff. Total length, 6·7 inches; wing, 3·6; tail, 1·2; tarsus, 0·95.

Range.—India and Northern Ceylon; extending eastward of the Bay of Bengal through Burma and Tenasserim to the Malay Peninsula, Siam, China, Formosa, and the Liu Kiu Islands.

Habits.—Mr. A. O. Hume writes:—"Scrub jungle, intermixed with patches of moderately high grass or dry ground, is perhaps the natural home of this species; but it may be met with anywhere in low bush jungle and on the skirts of forests, and in inhabited districts it greatly affects gardens, grass preserves, and similar enclosures. It strays into stubble and low crops in the mornings and evenings, even remaining in these at times throughout the day, but more generally retreating during the hotter noontide hours to the cover of some thorny bush or patch of grass upon their margins.

" Where the country is very arid, as in most parts of Rajputana and many places in the North-Western Provinces, this species is scarcely seen except during the rainy season; and again, it is almost unknown in densely cultivated and populated tracts where there is no jungle and no long grass. I have invariably seen it singly or in pairs, and only rarely in the latter; never in parties or bevies of five or six, as Jerdon says.

" Small millets, grass seeds, ants, white and black, and other small grains and insects constitute its food. It feeds almost exclusively in the early mornings and near sunset. At these times it may be seen running about along the paths of gardens or other enclosures, amongst isolated tufts of grass, on the margins of clumps of stunted jujube, and in the edges of low crops, and even in short stubbles, if these occur in the neighbourhood of suitable cover.

" It is a very silent bird, and except during the breeding season, I have never once heard it call; at that time the females emit a dull note, scarcely likely to attract attention unless you are on the look out for it. I have occasionally heard and noticed it, but not often.

" The most remarkable point in the life-history of these Bustard-Quails is the extraordinary fashion in which amongst them the position of the sexes is reversed. The females are the larger and handsomer birds. The females only call, the females only fight—natives say that they fight for the males, and probably this is true. What is certain is that, whereas in the case of almost all the other Game Birds it is the males alone that can be caught in spring-cages, &c., to which they are attracted by the calls of other males, and to which they come in view to fighting, in this species no male will ever come to a cage baited with a male, whereas every female within hearing rushes to a cage in which a female is confined, and if allowed to meet during the breeding-season, any two females will fight until one or other is dead or nearly so.

" The males, and the males only, as we have now proved in numberless cases, sit upon the eggs, the females meanwhile larking about, calling and fighting, without any care for their obedient mates ; and lastly, the males, and the males only, I believe, tend and are to be flushed along with the young brood.

" Almost throughout the higher sections of the animal kingdom you have the males fighting for the females, the females caring for the young; here, in one insignificant little group of tiny birds, you have the ladies fighting duels to preserve the chastity of their husbands, and these latter sitting meekly in the nursery and tending the young."

Colonel Butler writes :—" I found a nest containing four fresh eggs near Deesa on the 9th August. I laid a horse-hair noose on each side of the tuft of grass under which it was placed, and on returning to the spot about a quarter of an hour later, I found the *cock* bird snared and sitting upon the eggs, probably not knowing that he was caught, as he did not move off the eggs until I frightened him."

Nest.—A slight hollow in the ground lined with dry grass and sheltered by an overhanging tuft of grass, &c.

Eggs.—Three or four in number; pyriform; ground colour dirty white with pale lilac under-markings densely covered with brown and yellow specks and with some larger black blotches. Average measurements, 0·93 by 0·79 inch.

SUB-SP. A. THE ISLAND BUSTARD-QUAIL. TURNIX PUGNAX.

Hemipodius pugnax, Temm. Pig. et Gall. iii. pp. 612, 754 (1815).

Turnix pugnax, Blyth, Ibis. 1867, p. 309; Ogilvie-Grant, Cat. B. Brit. Mus. xxii. p. 534 (1893).

Adult Male.—Similar to male of *T. taigoor*.

Adult Female.—Chin and throat *black*, but differs chiefly from the female of *T. taigoor* in having a *fairly defined rufous collar* round the back of the neck contrasting with the colour of the back, in this respect approaching *T. fasciata* from the Philippine Islands. Total length, 6·6 inches; wing, 3·5; tail, 1·2; tarsus, 1.

Range.—South-Western Ceylon, Sumatra, Java, and Billiton.

II. THE PHILIPPINE BUSTARD-QUAIL. TURNIX FASCIATA.

Hemipodius fasciatus, Temm. Pig. et Gall. iii. pp. 634, 757 (1815).

Turnix nigrescens, Tweeddale, P. Z. S. 1877, p. 765.
Turnix haynaldi, Blasius, Ornis. iv. p. 317 (1888).
Turnix fasciata, Ogilvie-Grant, Cat. B. Brit. Mus. xxii. p. 535 (1893).

Adult Male.—Chin and throat *white,* narrowly edged with black *on the sides only ;* chest barred with black and *buff ; no* rufous nuchal collar *contrasting with* the colour of the back and upper parts, which resemble those of the female. Total length, 5·2 inches; wing, 3 ; tail, 1·1 ; tarsus, 0·9.

Adult Female.—Chin and throat *black ;* but may be distinguished from the females of the two previously-described forms by having *a well-defined rufous collar* round the back of the neck, *contrasting with* the colour of the back and rest of the upper-parts, which are mostly black, finely mottled with grey and mixed here and there with rufous. Total length, 6 inches ; wing, 3·4 ; tail, 1·1 ; tarsus, 1.

Range.—Philippine Islands.

III. THE CELEBEAN BUSTARD-QUAIL. TURNIX RUFILATUS.

Turnix fasciatus, Gould (*nec* Temm.), Birds of Asia, vii. pl. 11 (1861).
Turnix rufilatus, Wallace, P. Z. S. 1865, p. 480; Ogilvie-Grant, Cat. B. Brit. Mus. xxii. p. 536 (1893).

Adult Male.—Chin and throat *white,* narrowly edged with black *on the sides only ;* chest barred with black and *white ;* sides of the belly, thighs, and under tail-coverts *rufous.* Total length, 5·6 inches ; wing, 3·4 ; tail, 1·1 ; tarsus, 0·95.

Adult Female.—Chin and throat *white, barred with black,* like the breast ; sides of the belly, thighs, and under tail-coverts *rufous ;* general colour above warm brown, greyish on the mantle, and inclining to rufous on the rump, all finely mottled with black ; wings as in *T. taigoor.* Total length, 6·6 inches ; wing, 3·6 ; tail, 1·3 ; tarsus, 1·1.

Range.—Confined to the island of Celebes.

IV. THE SUMBAWA BUSTARD-QUAIL.   TURNIX POWELLI.

Turnix powelli, Guillemard, P. Z. S. 1885, p. 510, pl. xxix.;
Ogilvie-Grant, Cat. B. Brit. Mus. xxii. p. 537 (1893).

Adult Male.—As in the male of *T. rufilatus*, the chin and
throat white, *the sides only narrowly edged with black;* the chest
barred with black and *white.* Distinguished by the *absence*
of rufous on the belly, thighs, and under tail-coverts.

Adult Female.—Chin and throat *white, barred with black* like
the breast; but distinguished from the female of *T. rufilatus*
by the entire absence of rufous on the sides of the belly, thighs,
and under tail-coverts, which are whitish.

Range.—Gunong Api Island, Sumbawa.

B. *Middle of the breast not transversely barred with black;
throat never black; middle tail-feathers lengthened,
pointed and edged with white or buff; feathers of the
upper-parts edged with white or buff, giving the back a
scaly appearance.*

V. THE ANDALUSIAN BUSTARD-QUAIL.   TURNIX SYLVATICA.

Tetrao sylvaticus, Desfontaines, Mém. Ac. R. Sc. Paris, 1787,
p. 500, pl. xiii.
Tetrao andalusicus and *T. gibraltaricus*, Gmel. S. N. i. pt. ii. p.
766 (1788).
Turnix africanus, Bonn. Tabl. Encycl. Méth. i. p. 6 (1791).
Hemipodius tachydromus, Temm.; Gould, B. Europe, iv. p. 264
pl. (1837).
Hemipodius lunatus, Temm. Pig. et Gall. iii. pp. 629, 756
(1815).
Turnix sylvaticus, Dresser, B. Europe, vii. p. 249, pl. 494
(1876); Ogilvie-Grant, Cat. B. Brit. Mus. xxii. p. 537
(1893); Irby, Orn. Gibraltar, 2nd ed. p. 241, pl.
(1895).

Adult Male and Female—Distinguished   from   the   following species by having the shoulder-feathers margined with *white or whitish-grey ;* sides of the breast *pale buff,* contrasting strongly with the uniform rust-red centre, each feather with *a heart-shaped black spot* near the extremity; general colour above dull light red. Size larger.

In the female the nape is generally nearly uniform dull light red, while in the male the scale-like margins to the feathers extend to the back of the head.

Male : Total length, 6 inches ; wing, 3·1 ; tail, 1·5 ; tarsus 0·85.

Female : Total length, 7 inches; wing, 3·7; tail, 1·8; tarsus 0·9.

Range.—Southern Europe and North Africa.

Habits.—In the vicinity of Tangier the Andalusian Bustard-Quail appears to be both resident and migratory, those which migrate passing northwards during May and June, and returning in September and October.

Colonel Irby writes :—"Near Gibraltar this species is very local and nowhere plentiful, apparently less so than is really the case, for they are difficult birds to flush, and if put up once will rarely rise a second time. Scattered here and there, they chiefly frequent palmetto (*Chamærops humilis*) scrub, and appear to be most common near the coast, being more abundant to the east of Queen of Spain's Chair, especially about the Lomo del Rey and a place called Las Agusaderas. In their flight and habits, from what I could observe of them, they resemble the Indian Button-Quail (*T. dussumieri*).

"I have often seen them among the rough grass and bents close to the seashore, but always near palmetto, and one bird in particular for a long time frequented a patch of thick herbage near the mouth of the 'First River.' . . .

"The males of this species, and, I believe, of all the genus, are very much smaller than the females. This difference is so striking that the cazadores always declare that there are two

species.   I have at different times kept these little birds alive,
and sent them to England, and they are easily reconciled to
captivity, becoming very tame and confiding pets ; at times
they coo in a moaning way, whence their trivial Spanish name
of *torillo* or little bull.   They also have another single note,
much like that of the female Quail, but less loud."

Nest.—A slight hollow in the ground, scantily lined with dry
grass and sheltered by a bush, &c.

Eggs.—Four in number ; broad ovals ; ground colour, dirty
white, thickly spotted and blotched with reddish-brown, dark
brown, and greyish-lilac ; average measurement, 1·1 by 0·8
inch.

SUB-SP. A. SMITH'S BUSTARD-QUAIL.   TURNIX LEPURANA.

Ortygis lepurana, Smith, Rep. Exp. Centr. Afr. App. p. 55 (1836).
Hemipodius lepurana, Smith, Illustr. Zool. S. Afr pl. xvi. (1838).
Turnix lepurana, Ogilvie-Grant, Cat. B. Brit. Mus. xxii. p. 539
    (1893).

Adult Male and Female.—Differ from *T. sylvatica* only in being
smaller.

Male—Total length, 5 inches ; wing, 2·9 ; tail, 1·5 ; tarsus,
0·8.

Female—Total length, 5·7 inches ; wing, 3·2 ; tail, 1·7 ;
tarsus, 0·9.

Range.—Africa south of about 13° north latitude ; recently
obtained at Aden.

Habits.—No doubt the habits of this species are very similar
to those of *T. sylvatica*.   Mr. C. J. Andersson writes :—"This
species is not uncommon in Great Namaqua Land during the
rainy season ; but I have never found many of these birds together, and it is rarely that more than one of them is flushed
at a time.   Their favourite resorts are rank grassy spots in the
neighbourhood of temporary rain-pools and periodical water-
courses ; here they run about with great celerity, and, when

hard-pressed, lie so close as almost to allow themselves to be trodden on before they take wing, after which it is really impossible to flush them a second time.   They feed on insects and seeds."

Mr. Ayres says that this species is scarce in the Transvaal and inhabits the open veldt.   It is solitary in its habits, and is seldom if ever found on the corn lands with the common Quail. He never met with more than two together.

Eggs.—Like those of *T. sylvatica*, but smaller, and the markings finer and closer.   Measurements, 0·88 by 0·72 inch.

VI. THE LITTLE BUSTARD-QUAIL.   TURNIX DUSSUMIERI.

Hemipodius dussumieri, Temm. Pl. Col. v. pl. 454, fig. 2 (1828).
Hemipodius sykesi, Smith, Ill. Zool. S. Afr. ii. (see *H. lepurana*, pl. 16, footnote) (1838).
Turnix dussumieri, Gould, B. Asia, vii. pl. 10 (1869); Hume & Marshall, Game Birds of India, ii. p. 193, pl. (1879); Oates, ed. Hume's Nests and Eggs Ind. B. iii. p. 371 (1890); Ogilvie-Grant, Cat. B. Brit. Mus. xxii. p. 540 (1893).

Adult Male and Female.—Shoulder-feathers margined with *golden buff or straw-colour ;* feathers of the sides of the breast buff, each with a black or black and rufous *spot* near the extremity ; middle of the breast buff and *not much brighter than* the sides.   The male is generally paler and somewhat smaller (wing, 2·7 inches) than the female, in which the measurements are : total length, 5 inches; wing, 2·9; tail, 1·5; tarsus, 0·75.

Range.- -India, Pegu, Hainan and Formosa.

Habits.—This species, also known as the little " Button Quail," is a comparatively common and widely-distributed species, and is to a considerable extent migratory, visiting and breeding in the Himalaya and other parts of Northern and Western India, where it is not seen except during the breeding-season.

M. Hume says :—"In Upper India I have almost exclusively met with it in patches of low, dense grass, and most generally in patches of this nature situated in Dhak (*Butea frondosa*), or other thin bush or tree jungles. Occasionally I have flushed it from low crops and not unfrequently from belts of grass surrounding and dividing fields of these.

"It is hard to find without dogs, only rises when hard pressed, rises almost silently, sails away for a dozen yards like some large bee, and drops suddenly into some dense tuft of grass whence, as a rule, it makes no attempt to run, and where the dogs will often pounce upon it.

"I have once or twice seen it feeding in the early mornings in the little open spaces intervening between thinly-set tufts of grass, growing in lands which are flooded during the rains. During these latter I have seen them gliding like mice about the paths of my own and other gardens, where there was plenty of moderately-high fine grass. Two or three shot during the cold season had eaten only grass seeds, while two shot in my garden at Etáwah had fed almost exclusively on termites."

Colonel Butler says :—"The note of this species is remarkable, being a mixture of a 'purr' and a 'coo,' and when uttering it, the bird raises its feathers and turns and twists about much in the same way as an old cock pigeon. I have often watched them in the act of cooing within a few yards of me. If an old bird gets separated from one of its young ones, it is sure to commence making this peculiar noise."

Nest.—Lined with grass and placed in a slight depression in the ground in some standing crop or patch of grass. Mr. Hume states that occasionally he has heard of partially or wholly domed or covered-in nests being met with.

Eggs.—Usual number four, but five and even six are said to have been found—laid from April to October according to season and locality. The eggs are moderately broad ovals, much pointed towards one end, and fairly glossy, of a pale yellowish stone-coloured ground, minutely freckled all over with specks of yellowish and greyish-brown, overlaid with somewhat larger streaks, spots, and mottlings of dark earthy brown, varying in shade in different eggs. Small spots or clouds of small inky purple are usually scattered amidst the other markings." (*A. O. Hume*). Average measurements, 0.76 by 0.67 inch.

VII. THE AFRICAN BUSTARD-QUAIL. TURNIX NANA.

Hemipodius nanus, Sundevall, Œfv. Vet.-Akad. Förhandl. Stockh. 1850, p. 110.
Turnix nana, Ogilvie-Grant, Cat. B. Brit. Mus. xxii. p. 541 (1893).

Adult Male and Female.—Easily distinguished from the preceding species by having the feathers on the *sides of the breast transversely barred with black and white* at the extremity. The male is smaller (wing, 2.9 inches) and less brightly coloured than the female; the black and white bars on the sides of the breast are more extended, and not confined to the outer row of feathers. Total length of *female*, 5.8 inches; wing, 3.2; tail, 1.5; tarsus, 0.9.

Range.—Africa, south of about 10° S. lat. to the Great Karroo.

VIII. THE SOUTH AFRICAN BUSTARD-QUAIL. TURNIX HOTTENTOTTA.

Hemipodius hottentottus, Temm. Pig. et Gall. iii. pp. 636, 757 (1815.)
Turnix hottentotta, Ogilvie-Grant, Cat. B. Brit. Mus. xxii. p. 542 (1893).

Adult Female.—Much like *T. nana*, but most of the feathers of the *middle of the breast and belly having a round black spot near the extremity*. Total length, 6.6 inches; wing, 3.4; tail, 1.7; tarsus, 0.95.

I have not examined a *male* example in which the sex has been satisfactorily ascertained.

Range.—Extreme south of the African continent, south of the Great Karroo.

Habits.—The habits of this species do not apparently differ from those of other allied forms.   More than two are never seen together, and, as a rule, it is met with singly.   Grassy places and reeds in dry valleys are its favourite haunts, and, when flushed, it only flies a short distance before dropping again into cover, through which it instantly decamps, running with great rapidity.   It is resident in the extreme south of the African continent.

Nest.—" I have taken several nests of this bird ; one was on a rocky head (koppie) near Swellendam, and others on the Kuggeas.   I never saw one in a valley." (*W. Atmore.*)

Eggs.—Pyriform, and of the usual Hemipodian type.   *Five* in number.   (*W. Atmore.*)

b¹.   Feathers of the mantle and back practically uniform.

IX.   WHITEHEAD'S BUSTARD-QUAIL.   TURNIX WHITEHEADI. SP. N.
(*Plate XLI.*)

Adult Male.—Most like the *male* of *T. dussumieri* in size and markings, and the middle tail-feathers lengthened, pointed, and edged with buff; but the general colour of the upper-parts is *dark blackish-grey,* indistinctly washed on the back with rufous ; only a few feathers on the sides of the mantle and back have the outer-webs edged with buff, and, conse-quently, *no scaly appearance is produced,* the back being nearly uniform in colour.   The shoulder-feathers (scapulars) and secondary quills are widely edged with golden buff.   Total length, 4·4 inches; wing, 2·3 ; tail, 0·85 ; tarsus, 0·7.

In an immature bird, marked female, the tail is somewhat longer, measuring 1·05 inch.

Range.—Luzon, Philippine Islands.

PLATE XLI.

WHITEHEAD'S BUSTARD-QUAIL.

Mr. John Whitehead, in whose honour this species has
been named, has recently sent me two adult males and an
immature female of this species, obtained in the vicinity of
Manila. In most instances, one would hesitate to describe a
new species of *Turnix*, without having an adult female for
comparison, but the males before me are so totally distinct
from any species hitherto described, that there can, in this
instance at least, be no doubt as to its being a new form. It
is very curious that this species should not have been met with
by any of the numerous naturalists who have collected round
about Manila, and it only shows how easily birds of this kind
may be overlooked.

*d. Middle tail-feathers not lengthened and pointed, nor edged with
white or buff; feathers of the back without any scaly
appearance; shoulder-feathers not edged with golden-buff.*

X. BLANFORD'S BUSTARD-QUAIL. TURNIX BLANFORDI.

Turnix maculatus, Vieill. (*nec Hemipodius maculosus*, Temm.),
N. Dict. d'Hist. Nat. xxxv. p. 47 (1819).
Turnix blanfordi, Blyth, J. As. Soc. Beng. xxxii. p. 80 (1863);
Ogilvie-Grant, Cat. B. Brit. Mus. xxii. p. 542 (1893).
Turnix maculosa, Hume & Marshall, Game Birds Ind. ii.
p. 183, pl. (1879).
Hemipodius viciarius, Swinhoe, P. Z. S. 1871, p. 402 [adult
male].
Hemipodius chrysostomus, Swinhoe, Ann. Mag. N. H. (4) xii.
p. 375 (1873) [adult female].
Hemipodius variabilis, Prjevalski, Voy. Ussuri, no. 139.

Adult Male.—Differs from the *female* in having no rufous
nuchal collar. Total length, 5·8 inches; wing, 3·5; tail, 1·3;
tarsus, 0·95.

Adult Female.—General colour above greyish-brown, irregularly blotched and mottled with black and sometimes with
traces of rufous; *a well-defined rufous nuchal collar*; chin and
middle of throat whitish-buff; sides of throat, chest and breast

rufous-buff, shading into white on the belly ; most of the feathers *on the sides of the chest and breast with a round black spot near the extremity.* Total length, 6·5 inches ; wing, 3·8 ; tail, 1·5 ; tarsus, 1·05.

Younger examples of both sexes have the upper-parts blotched with black and mixed with rufous, the latter colour being most conspicuous on the back of the neck, where it forms an indistinct collar. Males may be recognised by their smaller size.

Range.—India, east of the Bay of Bengal to the south of Tenasserim, Siam and China, as far north as Manchuria.

Habits.—The late Mr. W. R. Davison says :—" I have always found this species about gardens or in the immediate vicinity of cultivation ; but it is very rare, being only occasionally met with, and always singly or in pairs. It is hard to flush, and only flies a short distance before again dropping, but it then runs a considerable distance before halting, and thereafter lies very close. It feeds like the other Quails in the mornings and evenings, lying hid during the heat of the day. On cloudy or rainy days it moves about all day.

"The fact is that it is apparently everywhere thinly distributed, that it is a terrible skulk, only to be flushed by chance without the aid of dogs, and is, I gather, as a rule, a very silent bird.

" Specimens examined had eaten grain, seeds, small insects, and tiny green shoots."

XI. THE INDIAN BUSTARD-QUAIL.   TURNIX TANKI.

Turnix tanki (Buchanan Hamilton), Blyth, J. As. Soc. Beng.
    1843, p. 180 ; Oates, ed. Hume's Nests and Eggs Ind. B.
    iii. p. 370 (1890); Ogilvie-Grant, Cat. B. Brit. Mus. xxii.
    p. 544 (1893).
Hemipodius joudera, Hodgson, in Gray's Zool. Misc. p. 85
    (1844) [nom. nud.].

Turnix joudera, Gray, Gen. B. iii. pl. cxxxi. (1846); Hume &
   Marshall, Game Birds Ind. ii. p. 187, pl. (1879).
Turnix dussumieri, Jerdon (*nec* Temm.), B. India, ii. p. 599
   (1863).

Adult Male.—Similar to the *female*, but the markings on the
upper-parts are coarser, and there is *no* rufous nuchal collar.
Total length, 5·3 inches; wing, 3; tail, 1; tarsus, 0·85.

Adult Female.—Like the *female* of *T. blanfordi*, but much
smaller; the back nearly uniform greyish-brown, with fine faint
wavy bars of darker brown; the rufous nuchal collar wider.
Total length, 5·5 inches; wing, 3·4; tail, 1·1; tarsus, 0·85.

Younger examples resemble immature specimens of *T. blanfordi*, but are, of course, smaller.

Range.—The Peninsula of India and east of the Bay of Bengal as far south as Tippera.

Habits.—Colonel Tickell says:—" This is a solitary bird, found
scattered about here and there throughout Bengal in open,
sandy, bushy places in and about jungles or fields and dry
meadows in cultivated country; frequently in low, gravelly
hills or uplands of 'Khunkur' (nodular limestone). It is met
with on both sides of the Ganges, at least as high up as
Benares."

Mr. A. O. Hume says:—" Its flight is even feebler and shorter
than that of the Bustard Quail (*T. taigoor*); it rises only when you
are about to step on it, with occasionally a low double chirp,
barely audible to my ears. When flushed it rises with much
less noise and whirr than do the Bustard Quails. It glides bee-
like through the air for a few paces, just skimming the waving
tops of the grass, and drops suddenly, as if paralysed, almost
before you can bring your gun to the shoulder.

"Smart little dogs will readily find it after it has thus drop-
ped, and as often as not (so pertinaciously does it cling to its
hiding-place) will seize it on the ground, but with only beaters

it is almost useless trying to put up one of these Button Quails a second time.

" Like all the Quails, they may be occasionally seen at early morn and eve feeding along the paths running through, or in tiny open spaces in the midst of, the grass they live in. I have never seen them in field or stubbles, nor had any of the few I have examined eaten any grain, only grass seeds and small black fragments, which might have been portions of small hard seeds or of tiny Coleoptera."

Nest.—Composed of soft blades of dry grass, placed at the foot of a tussock of coarse grass, the entrance-hole being on one side, and extending nearly to the top of the nest. (*E. A. Butler.*)

Eggs.—Smaller than those of *T. taigoor*, of a dirty yellowish white colour, thickly speckled, spotted and blotched all over with brownish-black. Shell highly glossed. Measurements, 0·84 by 0·63 inch. (*E. A. Butler.*)

XII. THE NICOBAR BUSTARD-QUAIL. TURNIX ALBIVENTER.

Turnix albiventris, Hume, Str. F. i. p. 310 (1873) ; Hume & Marshall, Game Birds Ind. ii. p. 199, pl. (1879) ; Ogilvie-Grant, Cat. B. Brit. Mus. xxii. p. 545 (1893).

Adult Male.—Like the *male* of *T. blanfordi*, but the back is darker brownish-grey, irregularly blotched and mottled with black and rufous. *No* rufous nuchal collar. Total length, 5·3 inches ; wing 3 ; tail, 1·1 ; tarsus, 0·85.

Adult Female.—Differs from the *female* of *T. blanfordi* in having the back darker brownish-grey, irregularly blotched and mottled with black and rufous ; the nuchal collar *deeper rufous and wider.* Total length, 5·5 inches ; wing, 3·2 ; tail, 1·2 ; tarsus, 0·9.

Younger examples resemble the immature of *T. blanfordi.*

Range.—Nicobar and Andaman Islands.

Habits.—The late Mr. W. R. Davison writes:—" This Quail is very rare in the Andamans, where I only once saw it, but at the Nicobars, at least on Camorta Island, it is not uncommon, frequenting the long grass, occasionally straying into gardens, &c. I have never seen them in coveys, but have found them usually in pairs, sometimes singly; they are difficult to get, as they will not rise without being almost trodden on. When they do rise, they only fly such a short distance that it would be impossible to fire without blowing them to pieces, and then they drop again into the long grass, from which it is almost impossible to flush them a second time. I have found them most numerous in the large grassy tracts in the interior of Camorta."

d¹. Shoulder-feathers edged with golden-buff.

XIII. TEMMINCK'S BUSTARD-QUAIL. TURNIX MACULOSA.

Hemipodius maculosus, Temm. Pig. et Gall. iii. pp. 631, 757 (1815).

Turnix maculatus, Vieill. Gal. des Ois. ii. p. 51, pl. 217 (1825) [adult female].

Hemipodius melanotus, Gould, B. Austr. v. pl. 84 (1848).

Turnix beccarii, Salvad. Ann. Mus. Civ. Genov. vii. p. 675 (1875).

Turnix maculosa, Ogilvie-Grant, Cat. B Brit. Mus. xxii. p. 546 (1893).

Adult Male.—Like the *female,* but there is no trace of a rufous nuchal collar. Total length, 5·1 inches; wing, 2·8; tail, 1·2; tarsus, 0·8.

Adult Female.—Like the *female* of *T. blanfordi,* but at once distinguished by the *golden-buff or straw-coloured edges to the shoulder-feathers* (scapulars); there is also more rufous in the plumage of the upper-parts below the rufous nuchal collar;

the throat and breast being *pale rufous*. Total length, 5·8 inches; wing, 3·2 ; tail, 1·3 ; tarsus, 0·85.

Range.—Northern and Eastern Australia and the interior. South coast of New Guinea, and South-East Celebes.

XIV. THE NEW BRITAIN BUSTARD-QUAIL. TURNIX SATURATA.

Turnix saturata, Forbes, Ibis. 1882, p. 428, pl. xii.; Ogilvie-Grant, Cat. B. Brit. Mus. xxii. p. 547 (1893).

Adult Male.—Like the *female*, but the chin and middle of of the throat are white, and the breast *rufous-buff*. Total length, 5·0 inches; wing, 2·9 ; tail, 1·0 ; tarsus, 0·8.

Adult Female.—Differs from the *female* of *T. maculosa* in having *no* rufous nuchal collar ; the whole of the upper-parts *blackish-grey*, with traces here and there of pale rufous and buff mottlings towards the tips of the feathers ; eyebrow-stripes, sides of the face, *throat*, and breast *bright rufous*.

Range.—New Britain and the Duke of York Archipelago.

Mr. Layard found this species mostly in the sweet-potato plantations on Mioko Island, Duke of York Archipelago.

Eggs.—Pyriform ; olive-brown, minutely speckled all over with tiny black or dark brown spots, sometimes forming blotches. Measurements, 1·0 by 0·95 inch.

XV. WALLACE'S BUSTARD-QUAIL. TURNIX RUFESCENS.

Turnix rufescens, Wallace, P. Z. S. 1863, p. 497 ; Ogilvie-Grant, Cat. B. Brit. Mus. xxii. p. 547 (1893).

Nearly Adult Male?—The only known example of this species is the type specimen in the British Museum.   The sex is not indicated, but it appears to be a nearly adult male. It resembles *T. maculosa* in having the general colour of the upper-parts greyish-brown ; but like *T. saturata* the breast is rufous.   It is impossible without additional specimens and adult females to say whether this species is really distinct from *T. maculosa*. Total length, 5·2 inches ; wing, 2·9 ; tail, 1·1 ; tarsus, 0·8.

Range.—Island of Semao, Timor.

C. *Neck and breast uniform bright rufous.*

XVI. THE CHESTNUT-BREASTED BUSTARD-QUAIL. TURNIX OCELLATA.

Oriolus ocellatus, Scopoli, Del. Flor. Faun. Insubr. pt. ii. p. 88 (1786).

Tetrao luzoniensis, Gmel. S. N. i. pt. ii. p. 767 (1788).

Hemipodius thoracicus, Temm. Pig. et Gall. iii. pp. 622, 755 (1815).

Turnix rufus, Vieillot, 2nd ed. Nouv. Dict. d'Hist. Nat. xxxv. p. 48 (1823).

Ortygis ocellata, Meyer, Nov. Act. Acad. C. L.–C. Nat. Curios. xvi. Suppl. i. p. 101, pl. 17 (1834).

Turnix ocellata, Ogilvie-Grant, Ibis. 1889, pp. 452, 469, pl. xiv.; id. Cat. B. Brit. Mus. xxii. p. 548 (1893).

Adult Male.—Differs from the female in having the chin and middle of the throat *white*, generally with some black spots ; *no* rufous nuchal collar ; the black ocelli on the wing-coverts larger and more numerous. Total length, 6·4 inches ; wing, 3·8 ; tail, 1·6 ; tarsus, 1·1.

Adult Female.—General colour above ashy-brown with irregular wavy black bars and mottlings, and some blotches of the same colour ; a fairly well-defined rufous nuchal collar ; wing-coverts brownish-buff, mostly with a black spot edged with whitish-buff near the tip of the outer web ; sides of the head and *throat black* (irregularly mixed with white in less mature examples) ; neck, chest, and breast *uniform rufous-chestnut ;* rest of under-parts dirty buff, paler on the belly. Total length, 6·9 inches; wing, 4·2 ; tail, 1·8 ; tarsus, 1·2.

Range.—Luzon, Philippine Islands.

II. *Leg (metatarsus) equal to or shorter than the middle toe and claw. Bill very stout in some species.*

XVII. THE MADAGASCAR BUSTARD-QUAIL. TURNIX NIGRICOLLIS.

Tetrao nigricollis, Gmel. S. N. i. pt. ii. p. 767 (1788).
Turnix nigricollis, Milne-Edwards & Grandidier, Hist. Nat. Madag. Ois. p. 494, pl. ccii. (1885) ; Ogilvie-Grant, Cat. B. Brit. Mus. xxii. p. 549 (1893).

Adult Male.—Differs chiefly from the *female* in having the feathers of the forehead black widely edged with buff ; the nape *like the upper back ;* the chin and throat *white ;* the sides of the chest washed with pale rufous ; the middle of the chest, breast, and flanks *buff, all barred with black;* belly paler. Total length, 5·5 inches ; wing, 2·9 ; tail, 1·35 ; tarsus, 0·8.

Adult Female.—The feathers of the forehead black, barred with white ; the nape *dark grey ;* the upper back mixed with black and rufous and margined with whitish-buff ; rest of upper-parts mostly brownish-grey, mottled with black and rufous ; the wing-coverts rufous with irregularly-shaped black and white spots ; the sides of the face and neck white, tipped with black; the chin, throat, and middle of the chest *black*, the two former bordered on either side by a white stripe ; the shoulders and sides of the chest *bright rust-red ;* the breast and belly *uniform dove-grey.* Bill slender. Total length, 5·8 inches ; wing, 3·3 ; tail, 1·4 ; tarsus, 0·8.

Range.—Madagascar.

XVIII. THE BLACK-BREASTED BUSTARD-QUAIL. TURNIX MELANOGASTER.

Hemipodius melanogaster, Gould, P. Z. S. 1837, p. 7 ; id. B. Austr. v. pl. 81 (1848).
Turnix melanogaster, Gould, Handb. B. Austr. ii. p. 178 (1865); North, Nests and Eggs B. Austr. p. 285 (1889); Ogilvie-Grant, Cat. B. Brit. Mus. xxii. p. 550 (1893).

Adult Male.—Differs from the *female* in having the *top of the head umber-brown* like the back; the sides of the head *white, tipped with black;* the chin and middle of the throat *pure white;* the chest and breast whitish-buff with a wide V-shaped black bar near the extremity of each feather. Total length, 6.3 inches; wing, 4.1; tail, 1.6; tarsus, 0.95.

Adult Female.—General colour above umber-brown, mixed with chestnut and barred with black; wing-coverts mostly chestnut with white black-edged spots; the forehead, sides of the face and throat *black;* top of the head mixed chestnut and black; chest and breast *black*, most of the feathers *tipped with white;* belly dark grey marbled with buff and black. Total length, 7.5 inches; wing, 4.4; tail, 1.7; tarsus, 1.0.

Range.—Eastern Australia.

Nest.—Merely a slight depression beside a tuft of grass.

Eggs.—Three to four in number; ground colour pale whitish-buff with pale lilac under-markings, minutely and thickly freckled all over with light reddish-brown and blotched with chestnut-brown and black. Measurements, 1.12 by 0.9 inch.

XIX. THE VARIEGATED BUSTARD-QUAIL. TURNIX VARIA.

Perdix varius, Latham, Ind. Orn. Suppl. p. lxii. (1801).
Hemipodius varius, Gould, B. Austr. v. pl. 82 (1848).
Hemipodius scintillans, Gould, P. Z. S. 1845, p. 62; id. B Austr. v. pl. 83 (1848).
Turnix varia, Gould, Handb. Austr. B. ii. p. 179 (1865);
    North, Nests and Eggs B. Austr. p. 285 (1889); Ogilvie-
    Grant, Cat. B. Brit. Mus. xxii. p. 551 (1893).

Adult Male.—Differs from the *female* in having *no* rufous nuchal collar; the chest buff, irregularly spotted and marked with grey. Total length, 6.4 inches; wing, 3.7; tail, 1.7; tarsus, 0.8.

Adult Female.—General colour above black, barred with rufous and shading into *chestnut on the back of the neck;* wing-coverts

with irregular white black-edged spots ; eyebrow-stripes, sides of
face and throat white, tipped with black, the chin and middle
of throat white ; feathers of the *chest grey*, with a spatulate buff
shaft-stripe ; breast buff, mixed with grey ; belly pale buff ; bill
fairly stout.   Total length, 7·6 inches ; wing, 4·2 ; tail, 1·9 ;
tarsus, 0.9.

Range.—Australia.   Recorded from New Caledonia, but the
bird from this island may prove to be distinct.

Habits.—Mr. Gould says :—" Among the Game Birds of Australia, the Varied Turnix plays a rather prominent part, for
although its flesh is not so good for the table as that of the little
Partridge and Quail (*Synoicus australis* and *Coturnix pectoralis*),
it is a bird which is not to be despised when the game-bag is emp-
tied at the end of a day's sport, for it forms an acceptable variety
to its contents.   Although it does not actually associate with
either of the birds mentioned above, it is often found in the
same districts, and all three species may be procured in the
course of a morning's walk in many parts of New South Wales,
Victoria, and South Australia, where it frequents sterile stony
ridges, interspersed with scrubby trees and moderately thick
grass.

" It is also very common in all parts of Tasmania suitable to
its habits, hills of moderate elevation and of a dry stony
character being the localities preferred ; it is also numerous on
the sandy and sterile islands in Bass's Straits.

" It runs very quickly, and when flushed flies low, its pointed
wings giving it much the appearance of a Snipe or Sandpiper.
When running or walking over the ground, the neck is
stretched out and the head carried very high, which, together
with the rounded contour of the back, give it a very
grotesque appearance.   The breeding-season commences in
August or September, and terminates in January, during which
period at least two broods are reared.

"The note is a loud and plaintive sound, which is often
repeated, particularly during the pairing-season.

"The young run as soon as they are hatched, and their appearance then assimilates so closely to that of the young Partridges and Quails that they can scarcely be distinguished.

" The food of this species consists of insects, grain, and berries; of the former many kinds are eaten, but locusts and grasshoppers form the principal part; a considerable quantity of sand is also found in the gizzard, which is very thick and muscular."

Nest.—A slight cavity, lined with dried grasses, close to a tuft of grass.

Eggs.—Four in number; wide ovals, slightly pointed at the smaller end, and marked much as in the other species, but the markings are generally very fine. Average measurement, 1·12 by 0·88 inch.

XX. THE CHESTNUT-BACKED BUSTARD-QUAIL. TURNIX CASTANONOTA.

Hemipodius castanotus, Gould, P. Z. S. 1839, p. 145; id. B. Austr. v. pl. 85 (1848).
Turnix castanonota, Ogilvie-Grant, Cat. B. Brit. Mus. xxii. p. 552 (1893).

Adult Male.—Like the adult *female*, but somewhat smaller.

Adult Female.—General colour above *uniform dull light* red, most of the feathers of the upper back with black and white edges, with some black blotches; wing-coverts ornamented with black and white ocelli; eyebrow-stripes and sides of the face *white, tipped with black;* chin and throat white; *chest* and breast *grey*, with white shaft-stripes; sides light red with irregular white black-edged ocelli; rest of under-parts whitish-buff. Bill *very stout*. Total length, 6 inches; wing, 3·6; tail, 1·5; tarsus, 0·9.

Range.—Northern Australia.

Habits.—Mr. Gilbert says:—"This species inhabits the sides of

stony hills in coveys of from fifteen to thirty in number ; when
disturbed, they seldom rise together, but run along the ground,
and it is only upon being very closely pursued that they will take
wing, and then they merely fly to a short distance.   While running along the ground their heads are thrown up as high as
their necks will permit, and their bodies being carried very
erect, a waddling motion is given to their gait, which is very
ludicrous.   The stomachs of those dissected were very muscular, and contained seeds and a large proportion of pebbles."

Eggs.—Differ from those of all the other species. Wide
ovals ; ground colour white, with comparatively very few
rounded black dots and spots, and a few greyish under markings.   Measurements, 1·05 by 0·8 inch.

XXI. THE RUFOUS-CHESTED BUSTARD-QUAIL.   TURNIX PYRRHOTHORAX.

Hemipodius pyrrhothorax, Gould, P. Z. S. 1840, p. 150 ; id. B.
    Austr. v. pl. 86 (1848).
Turnix pyrrhothorax, North, Nests and Eggs B. Austr. p. 287
    (1889) ; Ogilvie-Grant, Cat. B. Brit. Mus. xxii. p. 533
    (1893).
Turnix leucogaster, North, Ibis. 1895, p. 342.

Adult Male.—Resembles the adult *female*, but is smaller ; the
rust-coloured chest not so bright.   Total length, 5·2 inches ;
wing, 2·9 ; tail, 1·3 ; tarsus, 0·75.

Adult Female.—General colour above *stone-grey*, most of the
feathers of the back with narrow cross-bars of rufous and black ;
feathers of the back of the neck rufous-grey with whitish-buff
edges ; eyebrow-stripes, sides of face and neck white, tipped with
black ; *chest*, sides of breast and flanks *rufous ;* middle of the
throat and rest of under-parts whitish.   *Bill very stout.*   Total
length, 6 inches ; wing, 3·3 ; tail, 1·4 ; tarsus, 0·8.

Range.—North-East, East and South Australia, extending
westwards to the interior.

Nest.—A shallow hollow lined with dry grass.

Eggs.—Four in number; broad ovals; ground colour dull whitish, almost hidden by dense indistinct marking of chestnut and greyish-brown. Measurements, 1·0 by 0·77 inch.

XXII. THE SWIFT BUSTARD-QUAIL. TURNIX VELOX.

Hemipodius velox, Gould, P. Z. S. 1840, p. 150; id. B. Austr. v. pl. 87 (1848).

Turnix velox, Gould, Handb. Austr. B. ii. p. 184 (1865); North, Nests and Eggs B. Austr. p. 286 (1889); Ogilvie-Grant, Cat. B. Brit. Mus. xxii. p. 553 (1893).

Adult Male.—Similar to the adult *female*, but rather smaller. Total length, 5·5 inches; wing, 2·9; tail, 1·2; tarsus, 0·6.

Adult Female.—General colour above *dull bright red*, shading into light red on the nape and crown of the head; the markings very similar to those of *T. pyrrhothorax ; the sides of the head and chest pale light red ;* the breast and rest of underparts white. Bill *very stout.* Total length, 5·5 inches; wing, 3·3; tail, 1·2; tarsus, 0·7.

Range.—Australia.

Habits.—Mr. Gould says :—" It appears to give preference to low stony ridges thinly covered with grasses, for it was in such situations that I generally found it, though on some occasions I started it from among the rank herbage clothing the alluvial soil of the bottoms. It lies so close as to be nearly trodden upon before it will rise, and when flushed flies off with such extreme rapidity, as, when its small size and the intervention of trees combine, to render it a most difficult shot to the sportsman. On rising, it flies to the distance of one or two hundred yards within two or three feet of the surface, and then suddenly pitches to the ground. As might be expected, it lies well to a pointer, and it was by this means that I found many which I could not otherwise have started.

" It breeds in September and October."

12   U

Nest.—Slightly constructed of grasses placed in a shallow depression of the ground, under the shelter of a small tuft of grass. (*Gould.*)

Eggs.—Four in number; broad ovals; of a dirty white, either finely freckled all over or thickly blotched with markings of reddish-brown, light brown, and slate-grey. Measurements, 0·95 by 0·75 inch.

THE PLAIN-WANDERERS. GENUS PEDIONOMUS.

Pedionomus, Gould, P. Z. S. 1840, p. 114.

Type, *P. torquatus*, Gould.

Distinguished from the genus *Turnix* by possessing a small hind toe (hallux).

Only one species is known.

I. THE COLLARED PLAIN-WANDERER. PEDIONOMUS TORQUATUS.

Pedionomus torquatus, Gould, P. Z. S. 1840, p. 114; id. B
    Austr. v. pl. 80 (1848); Diggles, B. Austr. ii. pt. xv. pl.
    195 (1867); North, Nests and Eggs B. Austr. p. 288
    (1889); Ogilvie-Grant, Cat. B. Brit. Mus. xxii. p. 544
    (1893).
Pedionomus microurus, Gould, P. Z. S. 1842, p. 20.
Turnix gouldiana (Des Murs); Bonap. Compt. Rend. xlii. p.
    881 (1856).

(*Plate XLII.*)

Adult Male.—Differs from the *female* in having *no* rust-colour on the nape; the collar round the neck *buff* and brownish, not differing so conspicuously in colour from the rest of the plumage; the upper-chest washed with *bright buff*. Total length, 5·8 inches; wing, 3·4; tail, 1·2; tarsus, 0·9.

Adult Female.—General colour above brown, finely barred with black; a collar of *black white-tipped feathers* surrounding the neck; *nape and upper-part of chest rust-colour;* most of

PLATE XLII.

Wyman & Sons. Limited

COLLARED PLAIN WANDERER.

the feathers of the top of the head, back and shoulder-feathers margined with whitish-buff, and with the vanes free, giving the plumage a Rhea-like appearance; chin and middle of throat white; breast and rest of under-parts buff barred with black, except in the middle of the belly. Total length, 6·3 inches; wing, 4; tail, 1·6; tarsus, 1.

Range.—Australia; New South Wales, Victoria, South Australia, and the interior.

Habits.—Sir George Grey says:—"These birds are migratory; they appear at Adelaide in June, and disappear about January; where they go has not yet been ascertained. They never fly if they can avoid so doing, and are often caught by dogs; when disturbed they crouch down and endeavour to hide themselves in a tuft of grass. While running about they are in the habit of raising themselves in a nearly perpendicular position on the extremities of their toes, so that the hinder part of the foot does not touch the ground, and of taking a wide survey around them. . . . The call of those we have in confinement precisely resembles that of the Emu, not the whistle, but the hollow-sounding noise like that produced by tapping on a cask which the Emu utters, but it is, of course, much fainter."

October and November are said to be the principal breeding months.

Nest.—Made of dry grasses, and placed in a slight depression in the ground, under the shelter of a shrub or tuft of grass.

Eggs.—Four in number; pyriform; ground colour stone-white, thickly freckled and blotched with umber-brown and vinous-grey, the latter colour appearing as if beneath the surface of the shell. Average measurements, 1·35 by 0·94 inch.

APPENDIX TO VOL. I.

Page 42. LAGOPUS RUPESTRIS.

Professor D. G. Elliot has described a species from Attu Island, Alaska, as *Lagopus evermanni*, and another race from Kyska and Adak islands in the Aleutian chain as *L. rupestris townsendi* (Auk, xiii. pp. 24-29, 1896). The former is figured (t. c. pl. iii.), and appears to be indistinguishable from the male of the common Ptarmigan of Europe (*Lagopus mutus*) in breeding or summer plumage, a fact which seems strongly to favour the theory that only one polymorphic species of Ptarmigan exists.

Page 97, add :—

SPATZ'S RED-LEGGED PARTRIDGE.

Caccabis spatzi, Reichenow, J. f. O. 1895, p. 110.

Under the above name Dr. Reichenow has separated the paler desert form of the Barbary Red-legged Partridge met with in the south of Tunis from the darker birds inhabiting the wooded steppes of the north, which he regards as typical *C. petrosa*.

Caccabis petrosa, like *Caccabis chukar* (cf. vol. i. p. 93), might no doubt be divided into several races, the colour of the plumage being darker in wooded districts, where the annual rainfall is greater than in the more arid deserts. Generally speaking, it is useless to call such climatic races by different names, for they merge imperceptibly into one another, and it is impossible to define where one begins and the other ends.

Page 101, add :—

III. CHOLMLEY'S SEE-SEE PARTRIDGE.   AMMOPERDIX
CHOLMLEYI, sp. n.

Ammoperdix cholmleyi, Ogilvie-Grant, Handb. Game-Birds
(Allen's Nat. Libr.), ii. app. p. 293 (1896).

Adult Male.—The See-See inhabiting Palestine and the eastern shores of the Red Sea has always been considered identical with the African form met with in North-East Africa, in Egypt, and the countries bordering the western shores of the Red Sea. Until recently I had not examined a male of the African form, but my friend, Mr. A. J. Cholmley, during his recent trip to the Soudan, procured two fine males in the Erba Mountains, near Suakim.   On comparing these and two other African males recently added to the British Museum collection with the typical examples of *A. heyi* from Arabia, I find that the former differ constantly in having the general colour of the upper-parts darker, and in *lacking entirely* the white forehead and lores characteristic of *A. heyi*.   Measurements the same as those of *A. heyi*.

Adult Female.—Similar to the female of *A. heyi*.

Range.—Egypt and Nubia.

Page 119, add :—

XXI*a*. CRAWSHAY'S FRANCOLIN.   FRANCOLINUS CRAWSHAYI.

Francolinus crawshayi, Ogilvie-Grant, Ibis, 1896, p. 482, pl. xii.

Adult Male.—Most nearly allied to *F. levaillanti*, which it resembles in having the black and white superciliary stripes *confluent on the nape*.   It is easily distinguished by having the pure white chin and throat bordered by a dull rust-coloured band ; this is divided from the dull chestnut sides of the head and neck by a nearly pure white band, commencing above the angle of the gape ; the fore-neck is nearly pure white, only a few of the lowest feathers barred with black. Total length, 11 inches ; wing, 6·5 ; tail, 2·75 ; tarsus, 1·9.

Range.—Mountains of Nyika to the west of Lake Nyasa.

The only specimen I have seen was obtained by Mr. Richard Crawshay, on Cheni-Cheni Mountain, at an elevation of 7,400 feet.

Page 119, add:—

Francolinus kikurjuensis, Ogilvie-Grant, Bull. B. O. Club, v. p. xxiv. (1897).

Most nearly allied to *F. levaillanti*, but the middle of the throat is suffused with chestnut ; the feathers of the superciliary stripes and the stripes from the gape along the sides of the throat are pale rufous, with narrow black edgings, very different from the boldly marked black and white stripes in *F. levaillanti*. The patch of black and white feathers so conspicuous on the fore neck and upper part of the chest in *F. levaillanti* are represented by a much smaller patch, with the ground colour rufous and white   The breast and under parts are buff barred with black, especially on the sides and flanks ; the chestnut markings, so conspicuous in *F. levaillanti*, being at most merely represented by one or two scattered red spots on the outside flank feathers. Total length, about 12·0 inches ; culmen, 1·45 ; wing, 6·8 ; tail, 3·1 ; tarsus, 2·0.

Range.—Kikurju, British East Africa.

APPENDIX TO VOL. II.

Page 114, add :—

IV. DENDRORTYX HYPOSPODIUS.

Dendrortyx hypospodius, Salvin, Bull. B. O. Club, v. p. 5 (1896), allied to *D. leucophrys,* but with the under parts not spotted with rufous; breast and flanks dark grey, with a black shaft streak.

Range.—Costa Rica.

Page 127, add :—

IV. LOPHORTYX LEUCOPROSOPON.

Lophortyx leucoprosopon, Reichenow, Orn. Monatsber. iii. pp. 10, 97 with woodcut of ♂ and ♀ (1895).

Under the above title Dr. Reichenow describes and figures what he believes to be a new species of *Lophortyx.* The species is founded on birds bred in captivity, and the young birds are said to have been perfectly similar to their parents, but the origin of the latter is unknown.

ERRATUM.

On page 276 read :—

IX. WHITEHEAD'S BUSTARD-QUAIL. TURNIX WHITEHEADI.

Turnix whiteheadi, Ogilvie-Grant, Handb. Game-Birds (Allen's Nat. Libr.), ii. p. 276 (1896).

(*Plate XLI.*)

ALPHABETICAL INDEX.

Tetrao virginianus. 135.
Tetraogallus tibetanus. 20.
Texan Bob-White. 139, 140.
 Colin. 139.
 texanus, Colinus. 139.
 Ortyx. 139, 142.
Thaumalea. 44.
 amherstiæ. 47.
 obscura. 46.
 picta. 45.
Thick-billed Partridges. 151.
thoracicus, Dactylortyx. 150.
 Hemipodius. 282.
 Ortyx. 150.
Three-toed or Bustard-Quails. 267.
 Quails. 262.
tiarata, Numida. 94.
 Querelea. 94.
Tibetan Snow-cock. 20.
tibetanum, Polyplectron. 61, 64.
tibetanus, Pavo. 61.
 Tetraogallus. 20.
Tock'ro. 153.
tomentosa, Crax. 215.
 Mitua. 215.
 Pauxi. 215.
 Urax. 215.
Top-knot Quail, White. 115.
torquatus, Pedionomus. 289.
 Phasianus. 24, 26, 27, 28, 29,
  30, 39.
Touie. 204.
townsendi, Lagopus. 293.
trinkutensis, Megapodius. 165.
True Curassows. 200.
 Pheasants. 6.
tschudii, Chamæpetes. 256.
tuberosa, Crax. 214.
 Mitua. 214.
tumulus, Megapodius. 175.
Turkey, American. 106.
 Elliot's. 105.
 Florida. 108.
 Governor Battenberg's. 260.
 Honduras. 110.
 Mexican. 103.
Turkey, Wild. 107.
 Yacon. 253.
Turkey-like Guinea-Fowl. 86.
Turkeys, Brush. 163, 188.
Turnicidæ. 263.

Turnix africanus. 270.
 albiventris. 280.
 beccarii. 280.
 blanfordi. 277, 279, 280.
 castanonota. 287.
 dussumieri. 271, 273, 279.
 fasciata. 268, 269.
 gouldiana. 289.
 haynaldi. 269.
 hottentotta. 275.
 joudera. 278.
 lepurana. 272.
 leucogaster. 287.
 maculatus. 276, 280.
 maculosa. 276, 280, 281.
 melanogaster. 284.
 nana. 275.
 nigrescens. 269.
 nigricollis. 284.
 ocellata. 283.
 ocellatus. 265.
 plumbipes. 265.
 powelli. 270.
 pugnax. 268.
 pyrrhothorax. 288.
 rostrata. 265.
 rufescens. 282.
 rufilatus. 269, 270.
 rufus. 282.
 saturata. 282.
 sylvatica. 270, 272.
 taigoor. 265, 278.
 tanki. 278.
 Varied. 285.
 varius. 285.
 velox. 289.
 whiteheadi. 276.

Uatáraku. 243.
unicolor, Chamæpetes. 257.
Urax tomentosa. 215
 urumutum. 212.
 Crax. 211.
 Nothocrax. 211, 212.
 Urax. 212.
urvilii, Alecthelia. 179.

validirostris, Crax. 211.
Valley Partridge. 122.
vallicola, Callipepla. 121.

www.ingramcontent.com/pod-product-compliance
Lightning Source LLC
Chambersburg PA
CBHW021357210326
41599CB00011B/916

9783744734431